ENGINEERING TOOLS, TECHNIQUES AND TABLES

CASE-BASED REASONING: PROCESSES, SUITABILITY AND APPLICATIONS

ENGINEERING TOOLS, TECHNIQUES AND TABLES

Additional books in this series can be found on Nova's website under the Series tab.

Additional E-books in this series can be found on Nova's website under the E-books tab.

ENGINEERING TOOLS, TECHNIQUES AND TABLES

CASE-BASED REASONING: PROCESSES, SUITABILITY AND APPLICATIONS

ANTONIA M. LEELAND
EDITOR

Nova Science Publishers, Inc.
New York

For permission to use material from this book please contact us:
Telephone 631-231-7269; Fax 631-231-8175
Web Site: http://www.novapublishers.com

LIBRARY OF CONGRESS CATALOGING-IN-PUBLICATION DATA

Case-based reasoning : processes, suitability and applications / editors,
Antonia M. Leeland.
p. cm.
Includes index.
ISBN 978-1-61728-352-9 (hardcover)
1. Case-based reasoning. I. Leeland, Antonia M.
Q338.8.C39 2009
153.4'3--dc22
2010025433

Published by Nova Science Publishers, Inc. † New York

CONTENTS

PREFACE

Chapter 1 - A popular approach in Artificial Intelligence involves integration or combination of (two or more) representation methods. The integrated components offer advantages to the overall system. Integrated approaches have been applied to various application domains demonstrating their effectiveness in knowledge representation and reasoning. Integrations of case-based reasoning with other intelligent methods have been explored deriving effective knowledge representation schemes. Case-based reasoning is usually combined with rule-based reasoning, model-based reasoning and soft computing methods (i.e., fuzzy methods, neural networks, genetic algorithms). Certain types of case-based reasoning integrations have been extensively explored. However, other types of combinations have not been adequately investigated, which leaves room for extensive research work. In this chapter, the authors illustrate basic types of case-based reasoning integrations. A categorization scheme for such integrations is provided and the functionality of specific approaches combining case-based reasoning with other intelligent methods is presented. The focus is on integrations dealing with innovative ideas and representing research areas that need to be explored. The chapter also outlines a formalism combining case-based reasoning with neurules, a type of hybrid rules integrating symbolic rules with neurocomputing. Moreover, future directions are pointed out.

Chapter 2 – Conventional statistic and artificial intelligence approaches to business prediction resolve classification and estimation tasks in accordance with supervised learning. To improve prediction accuracy of supervised learning approaches, a priori information such as discrete outcomes of classification is required. In a domain dealing with practical business problem however, the priori information resulting from the classification task is not generally pre-defined or unknown.

Business prediction research studies of statistical and artificial intelligence based methodologies in a domain of corporate failure prediction clearly defining the classification outcomes by whether the corporation is bankrupt or not have been widely studied. More recently Case-Based Reasoning (CBR), which extracts the most relevant case to predict the credit level of the targeted corporation, is being increasingly denoted as an alternative approach.

Although CBR is applied to an environment where the adequate and appropriate classification outcomes are not available, studies finding out generalized indexing and retrieving patterns from supervised learning with abundant data have been carried out for performance improvement of CBR. However, those approaches can be restricted in a practical situation as the underlying foundation of CBR originates from a problem-solving paradigm using case specific knowledge of past experience to solve a new problem. To overcome this limitation, the authors propose a hybrid approach combining CBR and Data Envelopment Analysis (DEA).

The results demonstrate that the hybrid approach combining CBR and DEA suggested an alternative methodology to weigh features with limited priori information. Further, through the comparative experiment between Multiple Discriminant Analysis (MDA) as supervised learning and their hybrid approach as unsupervised learning, the results indicate the hybrid approach performs better in bankruptcy prediction than MDA when priori information is unavailable. Additionally, key areas influencing CBR's prediction performance including weighting features, determining case base, selecting an optimal number of combining cases and comparing distance metrics are considered.

Chapter 3 – This paper reviews the Case-Based Reasoning (CBR) approach which, over the last few years, has grown from a rather specific and isolated research area into a field of widespread interest both from academic and commercial standpoints, and has been developed into a theory of problem-solving and learning for computers and people.

More explicitly, following an introduction with the basic concepts and a brief historical background of CBR, the authors focus on the steps of the CBR process, the several types of the CBR methods, the applications of CBR to a wide range of domains, and on the development trends of methods, applications and research for CBR. Finally, in their conclusion section, the authors underline the differences between CBR and the classical rule-induction algorithms, the authors refer to the existing criticism for CBR methods and, summarizing the paper, the authors derive their final conclusion about the CBR approach.

Chapter 4 – Given that customers can reserve many days before the service day, they, not uncommon, are also allowed to cancel their reservations before using the service. Expecting the volume of cancellations helps operators effectively manipulate their resources, which in many cases are limited and also perishable such as hotel rooms and railway seats. This paper proposes a case-based predicting model for the purpose of cancellation forecasting. Under the stages of retrieving, reusing and revising, the first major contribution of this paper is on the inclusion of temporal features of curves in the stage of pattern retrieval. Temporal features such as day-of-week, recency effect and reliability of information over booking days are investigated to generate a reasonable method for case retrieval. Another contribution is the integration of a direct search algorithm for parameter estimation in the stage of revising. Hooke-Jeeves algorithm is applied to search five key parameters in the proposed predicting model. The empirical study, which uses real railway data, shows that the proposed case-based predicting model can have at least 20% improvement of MSE over pick-up and regression models, which are two popular benchmarks in practice. Similar concepts can be extended to other industries with cancellation behavior such as airlines, restaurants, hotels, rental cars, golf courses etc.

Chapter 5 –The problem statement, and the process of solving the problem, bound up with the investigation of safety related to complex technological systems on the basis of the method of case-based reasoning is considered. Case-based expert systems provide decision support on the principle of analogy and on the basis of experience data available.

Conceptualization and formalization of data and knowledge, which represent the state of technological systems, have been conducted. Principal properties and hazardous states of the technological systems have been identified. Hazardous states are represented in the form of a cause-effect sequence of states that are characterized by the increased risk level: failure, accident, emergency, and technogenic catastrophe. An object-oriented model of a case, which describes the proposed dynamics of hazardous states of some technological object on account of the complex technological system's structure and the hierarchy of hazardous states, has been developed. The elaborated case-base contains information about 200 incidents and failures, which have taken place at the USSR and Russian chemical and petrochemical enterprises.

The algorithms and the software system, which provide for finding the solution in the hierarchical space of cases, as well as for adaptation of the solution on the basis of production rules and analytical models, have been elaborated.

At the end of elevating the efficiency of the case retrieval procedure the authors have conducted the indexing of cases by using elements of the object-oriented model. The software system provides for decision support in the processes of providing for the safety of technological systems and compensating for the consequences of failure, while including the solving of problems related to identifying hazardous states, determining their causes, assessing the degree of hazard and forecasting scenarios of the evolution of hazardous states.

A real example, demonstrating the decision support in the process of determination of the causes of failure, which is provided for by the software proposed, is considered.

Chapter 6 – In this chapter the authors introduce a finite absorbing Markov chain having as states the main steps of the Case-Based Reasoning (CBR) Process (retrieval, reuse, revision, and retaining), where retaining is the unique absorbing state. Applying standard results of the theory of finite Markov chains, the authors succeed in calculating the probabilities for the CBR process to be in a certain step at a certain phase of the solution of the corresponding real-world problem, and the authors obtain a measure for the effectiveness of the corresponding CBR system in solving similar new problems.

Next the authors present the first three of the above steps of the CBR process as fuzzy subsets in the set U of the linguistic labels of negligible, low, intermediate, high and complete success for each of the above steps. In this way, the authors build a fuzzy model for the representation of a CBR system, where the authors use the total possibilistic uncertainty as a measurement tool for its effectiveness in solving new related problems. Examples are also given to illustrate their results.

Chapter 7 – In Case-Based Reasoning, usually former single cases are considered. For medical diagnosis, the authors propose a method that does not retrieve single but generalised, prototypical cases (prototypes).

Since diagnosis of dysmorphic syndromes is a domain with incomplete knowledge and where even experts have seen only few syndromes themselves during their lifetime, documentation of cases and the use of case-oriented techniques are popular. So, most of the systems dealing with the diagnosis of dysmorphic syndromes, perform classification based on prototypes. Different prototypicality measures are applied to determine the most probable syndrome. These measures differ from the usual Case-Based Reasoning similarity measures, because here cases and syndromes are not represented as attribute value pairs but as long lists of symptoms, and because query cases are not compared with cases but with prototypes.

In contrast to other dysmorphic systems their approach additionally applies adaptation rules. These rules do not only consider single symptoms but combinations of them, which indicate high or low probabilities of specific syndromes. However, it is a big problem to acquire such domain dependent adaptation rules. First, some medical experts, with difficulties, provided a few rules. Recently, the authors have generated suggestions for adaptation rules automatically and have discussed them subsequently with medical experts.

Chapter 8 – Case-Based Reasoning (CBR) allows us to resolve problems in a dynamic environment, and propose a solution that follows a checking step, in which the authors proceed along the test-error-correction cycle until the authors reach the result aspired to. This paper proposes a novel model for CBR (the 3R model), in which Retrieve, Reuse, and Retain are the main tasks for the CBR process. The integration of mathematical reasoning allows the search for the weights that are assigned to the different attributes of a case, then, to come up with the wished for result, based on the similar case in one go. Hence, the classical 4R model is reduced through the elision of the Revise step, the results are then reached automatically (the search of the weights, the similar case and the final result). This model is used as a negotiation strategy to predict the seller's behaviour. The authors have applied it to the real estate domain and the authors have come up with interesting results.

Commentary - Case-Based Reasoning (CBR), as a general problem-solving methodology intended to cover a wide range of real-world applications, must face the challenge of dealing with uncertain, incomplete and vague information.

In: Case-Based Reasoning: Processes...
Editors: Antonia M. Leeland
ISBN: 978-1-61728-352-9

Chapter 1

CASE-BASED REASONING INTEGRATIONS: APPROACHES AND APPLICATIONS

Jim Prentzas[a,*b] ***and Ioannis Hatzilygeroudis***[a,**]
[a] University of Patras, School of Engineering Department of Computer Engineering & Informatics, 26500 Patras, Greece.
[b] Democritus University of Thrace, School of Education Sciences Department of Education Sciences in Pre-School
Age, 68100 Nea Chili, Alexandroupolis, Greece.

ABSTRACT

A popular approach in Artificial Intelligence involves integration or combination of (two or more) representation methods. The integrated components offer advantages to the overall system. Integrated approaches have been applied to various application domains demonstrating their effectiveness in knowledge representation and reasoning. Integrations of case-based reasoning with other intelligent methods have been explored deriving effective knowledge representation schemes. Case-based reasoning is usually combined with rule-based reasoning, model-based reasoning and soft computing methods (i.e., fuzzy methods, neural networks, genetic algorithms). Certain types of case-based reasoning integrations have been extensively explored. However, other types of combinations have not been adequately investigated, which leaves room for extensive research work. In this chapter, we illustrate basic types of case-based reasoning integrations. A categorization scheme for such integrations is provided and the functionality of specific approaches combining case-based reasoning with other intelligent methods is presented. The focus is on integrations dealing with innovative ideas and representing research areas that need to be explored. The chapter also outlines a formalism combining case-based reasoning with neurules, a type of hybrid rules integrating symbolic rules with neurocomputing. Moreover, future directions are pointed out.

---

* Email: dprentza@psed.duth.gr.
** Corresponding author: Email: ihatz@ceid.upatras.gr.

Keywords: case-based reasoning integrations, hybrid case-based reasoning, case-based reasoning combinations, hybrid intelligent systems, integrated intelligent systems, hybrid knowledge representation and reasoning, case-based reasoning.

1. Introduction

The combination or integration of (two or more) different problem solving and knowledge representation methods has proven effective in many application areas [49]. The aim is to create combined formalisms that benefit from each of their components. Disadvantages or limitations of specific intelligent methods can be surpassed or alleviated by their combination with other methods. It is worthwhile to explore combinations of different intelligent methods in case their advantages and disadvantages prove to be complementary to an adequate degree. Popular integrations are neuro-symbolic approaches, combining symbolic representations with neural networks [7], [29], neuro-fuzzy approaches, combining fuzzy logic and neural networks [52], approaches combining neural networks with genetic algorithms [2], approaches combining fuzzy or neuro-fuzzy systems with genetic algorithms [2] and approaches combining case-based reasoning with other intelligent methods [46], [47], [6], [61], [63]. Other integrations have been developed as well.

Case-based reasoning (CBR) exploits stored past cases whenever a similar new case needs to be dealt with [1], [38], [39], [18]. Cased-based inference is performed in four phases known as the CBR cycle [1]: (i) *retrieve*, (ii) *reuse*, (iii) *revise* and (iv) *retain*. The retrieval phase retrieves from the case base the most relevant stored case(s) to the new case. In the reuse phase, a solution for the new case is created based on the retrieved most relevant case(s). The revise phase validates the correctness of the proposed solution, perhaps with the intervention of the user. Finally, the retain phase decides whether the knowledge learned from the solution of the new case is important enough to be incorporated into the system. CBR is a useful approach in domains with a sufficient number of available (or obtainable) cases and does not require existence of an explicit domain model.

Integrations of CBR with other intelligent methods have been pursued in various domains. In such combinations, the combined system offers advantages in knowledge representation and reasoning compared to each of the combined methods working alone. CBR has been integrated with intelligent methods such as rule-based reasoning (RBR), model-based reasoning (MBR), fuzzy methods, neural networks, probabilistic reasoning, genetic algorithms and other methods as well.

When two or more intelligent methods are combined, different integration models can be employed [49]. Not all types of combination models have been employed in CBR integrations. An aspect of interest involves pointing out trends in CBR integrations in which there is room for extensive research work. A trend that needs to be explored further concerns approaches in which the problem solving process can be decomposed into subprocesses (tasks or stages) for which different representation formalisms are required or available. In such situations, a CBR system as a whole (with its possible internal modules) is integrated 'externally' with other intelligent systems in order to create an improved overall system. An interesting aspect of this combination trend is that different types of such combinations can be developed. This trend has been explored thoroughly for integrations of CBR with RBR and

MBR but not for integrations of CBR with other methods. Another trend that could also produce fruitful results involves approaches in which CBR is embedded within another intelligent method. Such approaches have been explored in integrations of CBR with genetic algorithms. However, they could prove to be effective in integrations of CBR with other intelligent methods as well. Moreover, combinations of CBR with certain specific intelligent methods have not been explored extensively. Such intelligent methods involve for instance the various neuro-symbolic approaches.

Due to the fact that several approaches integrating CBR with other intelligent methods have been developed, it is necessary to discuss issues involving main trends in such combinations that have been applied. In this discussion it is also necessary to point out interesting open aspects for future work. In this chapter, we discuss various aspects involving CBR integrations. We focus on key aspects involving CBR integrations and discuss the potential for future research work. We also briefly present an approach combining CBR with neurules, a neuro-symbolic knowledge representation scheme. Neurules are a type of hybrid rules integrating symbolic rules with neurocomputing [25], [26] and exhibit certain attractive features such as naturalness and modularity. Such an approach integrates three intelligent methods: symbolic rules, neural networks and CBR [28].

The purpose of the discussion included in this chapter is threefold. We believe that it will increase understanding of the field concerning integrations of CBR with other intelligent methods. In addition, it may lead to development of new (or overlooked) ways of combining CBR with other intelligent methods. Finally, it is a useful guide to developers/designers of such systems.

The structure of the chapter is as follows. Section 2 discusses issues involving main trends in CBR integrations. This discussion serves as background knowledge for the following sections. Section 3 briefly presents representative approaches of specific types of CBR integrations that could provide impetus for future research work. In section 4, we present an outline of an approach combining CBR with neurules. Finally section 5 concludes.

2. Trends in Integrations of CBR with Other Intelligent Methods

Various CBR integrations have been developed [63], [61], [46], [47]. To develop such integrations, existence of (or ability to acquire/construct) necessary knowledge sources corresponding to each of the combined methods is required. Other criteria may also be specified to judge whether an approach combining CBR with other intelligent method(s) could be applied to a specific domain [63].

To categorize CBR combinations one could use Medsker's general categorization scheme for integrated intelligent systems [49]. Medsker distinguishes five main combination models: *standalone*, *transformational*, *loose coupling*, *tight coupling* and *fully integrated models*. Distinction between those models is based on the degree of coupling between the integrated components.

In [61] Medsker's categorization scheme was extended and revised to accommodate recent advances in integrations of CBR with RBR. This new scheme provides a more consistent view to modeling integrations of CBR with other intelligent methods. Figure 1 depicts the categorization scheme for CBR integrations, based on that in [61]. For each (sub)category, intelligent method(s) with which CBR has been combined is shown besides each (sub)category. An unexplored type of CBR integration is indicated by a broken rectangle. In Figure 1, 'GA' stands for 'genetic algorithm' and 'NS' for 'neuro-symbolic approaches'. It should be mentioned that in [61] deficiencies of other categorization schemes for CBR integrations (e.g. [46], [47], [23]) are discussed.

Two main categories of CBR integrations are discerned in our categorization scheme: (a) standalone and (b) coupling approaches. Three main types of coupling approaches can be distinguished: (i) sequential processing, (ii) co-processing and (iii) embedded processing.

In standalone models, independent components of each approach are developed that do not interact with each other during reasoning. They can be used in parallel to compare the independent solutions providing an opportunity to compare the capabilities of each approach.

In sequential processing, the flow of information (produced by reasoning) between the integrated modules is sequential or semi-sequential. It includes approaches in which information necessarily passes sequentially through some or all of the combined components in order to produce the final result. Two subcategories of the sequential category are distinguished: the 'loosely coupled sequence' and the 'tightly coupled sequence' subcategories. The former involves approaches in which the output of one component does not play an important role in the internal reasoning process of the next component. The latter concerns approaches in which the output of one component plays a significant role in the internal reasoning process of the next component.

The tightly coupled subcategory is distinguished into two subcategories: compulsory sequence and conditional sequence. In compulsory sequence, a component is invoked unconditionally after the previous component in the sequence. In conditional sequence, the second component is invoked if the first one fails to provide a solution. All approaches belonging to loosely coupled sequence follow the conditional sequence pattern. An aspect of interest in sequential processing concerns the invocation order of the integrated components and more specifically, whether CBR is invoked before or after the other integrated components. In all existing sequence approaches but the tightly coupled conditional sequence approaches, CBR is invoked before or after invocation of other combined component(s). In existing tightly coupled conditional sequence approaches, CBR is invoked after the other integrated component(s).

In co-processing, the integrated components closely interact during reasoning. To produce output, flow of data between the components is bidirectional enabling an enhanced form of interaction. The integrated components may be also invoked in parallel to solve the problem. Approaches belonging to the co-processing category are distinguished to cooperation oriented and reconciliation oriented according to whether emphasis is given to cooperation or reconciliation respectively. In cooperation oriented approaches, the integrated components cooperate with each other during inference. In reconciliation oriented approaches, a reconciliation process is necessary since each integrated component produces its own conclusion, possibly differing from the conclusion of the other component. Cooperation oriented approaches may either employ explicit reasoning control or implicit reasoning control. The former approaches employ an explicit controller or explicit control knowledge to coordinate reasoning. The latter approaches coordinate reasoning implicitly.

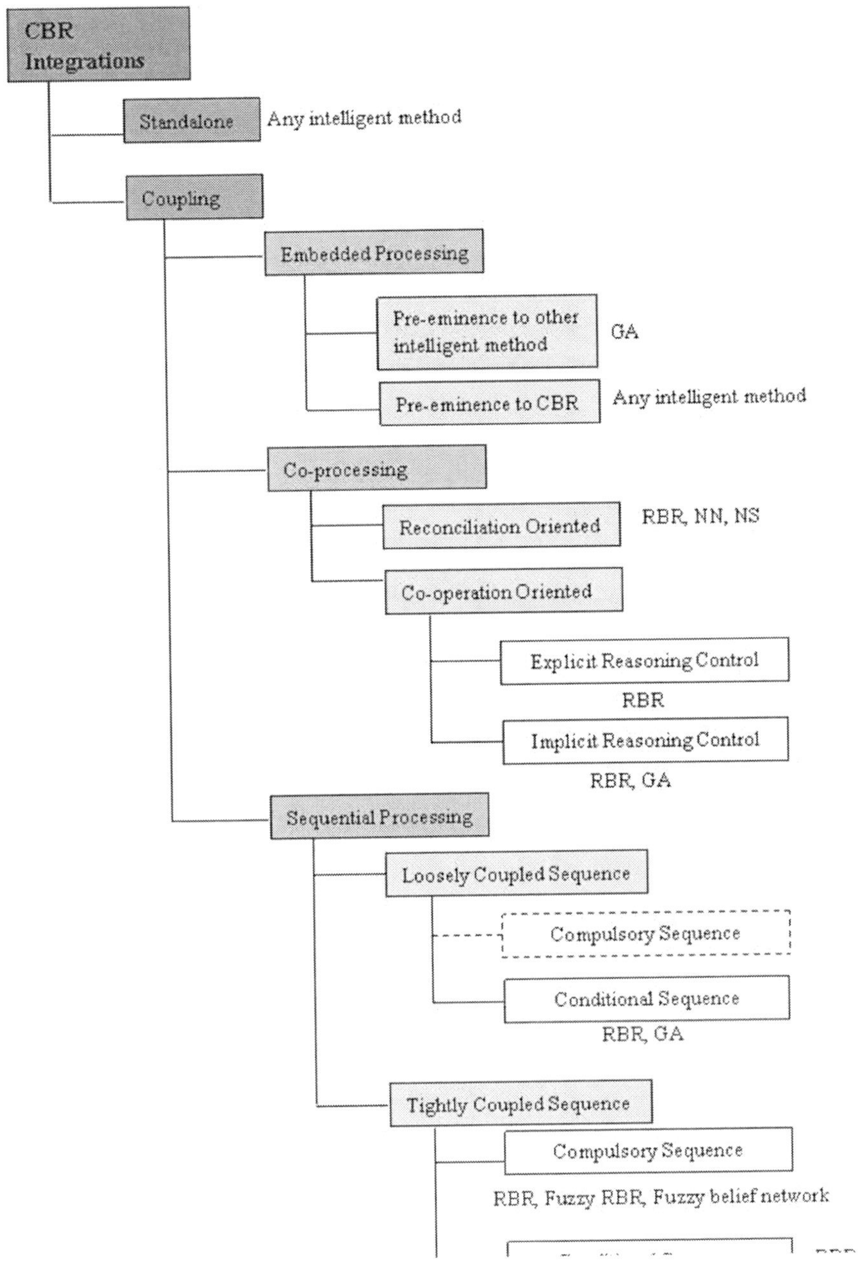

Figure 1. Categorization scheme for CBR integrations.

In embedded processing, a component based on one approach is the primary problem solver, embedding component(s) based on other representation method(s) to handle its internal reasoning tasks. Embedded processing approaches can be distinguished into those giving pre-eminence to CBR and to those giving pre-eminence to other method. In the former, a CBR system embeds other intelligent method(s) to assist various internal CBR tasks. Internal CBR tasks can be implemented using various techniques [73], [14], [53]. The latter involve the reverse (and less usual) approach i.e. embedding CBR within other representations to assist in their internal tasks.

Not all of these combination models and/or their underlying categories have been thoroughly explored in combinations of CBR with other intelligent methods. Obviously, the standalone model can be applied in combinations of CBR with any other intelligent method. In combinations of CBR with certain methods (e.g. RBR, MBR), various coupling approaches have been investigated [61], [46], [47]. However, in coupling combinations of CBR with soft computing methods, embedded approaches seem to be the most thoroughly investigated. Embedded coupling approaches mainly concern those giving pre-eminence to CBR. Embedded coupling approaches giving pre-eminence to other intelligent method do not seem to be popular with the exception of genetic algorithms (see Section 3.3).

Combinations of CBR with other intelligent methods can offer advantages to the overall system especially in case the advantages and disadvantages of the combined methods are to a certain degree complementary. CBR provides advantages to the overall system such as easy knowledge acquisition by exploiting available (or obtainable) cases, naturalness, modularity, incremental learning and certain explanation facilities. Other intelligent methods when combined with CBR may offer advantages to the overall system such as the following:

Table 1. Application domains and intelligent methods CBR has been integrated with.

Application Domain	Intelligent Method(s) CBR has been integrated with
Agriculture	RBR
Aircraft Design	RBR
Aircraft Fleet Maintenance	RBR
Automobile Construction	RBR
Banking	RBR
Biomedicine	RBR
Construction	RBR
Design of Nutrition Menus	RBR
E-learning, Intelligent Tutoring	RBR, GA
Emergency Fire Management	GA
Environmental Impact Assessment	Fuzzy RBR
Equipment Failure Analysis	RBR
Finance	RBR, Possibilistic RBR
Legal Reasoning	RBR
Life Insurance	RBR
Medicine	RBR, Fuzzy RBR
Modeling Event-based Dynamic Situations	RBR
Music	RBR
Personnel Performance Evaluation	RBR
Quality of Service	RBR
Real-Time Marine Environment Monitoring	RBR
Situation and Threat Assessment of Ground Battlespaces	Fuzzy belief network
Surname Pronunciation	RBR
Ultrasonic Rail Inspection	RBR

- RBR provides general and compact available domain knowledge in the form of rules and rule-based explanation facilities.
- Fuzzy methods provide imprecision handling and (in case of fuzzy RBR) fuzzy rule-based domain knowledge.
- Neural networks provide robustness, generalization, learning capabilities, classification/clustering capabilities.
- Genetic algorithms provide search and optimization facilities, compact representation of problem parameters and representation of possible solutions.
- Neuro-symbolic approaches provide (more or less) the combined advantages of symbolic methods and neural networks.
- Neuro-fuzzy approaches provide (more or less) the combined advantages of fuzzy methods and neural networks.

Table 2. Application domains and systems integrating CBR with other method.

Application Domain	Integrated Approaches
Agriculture	[78]
Aircraft Fleet Maintenance	[75]
Banking	[41]
Biomedicine	[55]
Construction	[20]
Design of Nutrition Menus	[45]
E-learning, Intelligent Tutoring	[31]
Emergency Fire Management	[8]
Environmental Impact Assessment	[43]
Equipment Failure Analysis	[33]
Finance	[16], [19]
Legal Reasoning	[64], [10], [11], [77], [12]
Life Insurance	[40]
Medicine	[9], [48], [51], [58], [65], [21]
Modeling Event-based Dynamic Situations	[34]
Music	[66]
Personnel Performance Evaluation	[17]
Quality of Service	[24]
Real-Time Marine Environment Monitoring	[71]
Situation and Threat Assessment of Ground Battlespaces	[44]
Surname Pronunciation	[23]
Ultrasonic Rail Inspection	[35]

Tables 1 and 2 summarize the application domains in which non-embedded CBR combinations have been developed. For each domain, Table 1 depicts the intelligent method(s) CBR has been integrated with. Table 2 depicts specific systems for each domain. It should be mentioned that some of the systems depicted in Table 2 whose application domain does not strictly concern e-learning have been employed as teaching assistants. Such systems are presented in [9], [21]. Moreover [78] is also reported that could be used as a teaching assistant. It should be mentioned that several integrated approaches do not involve specific application domains and their effectiveness has been tested with datasets.

Generally speaking, the following unexplored research directions regarding CBR integrations can be discerned:

- Implementation of CBR combinations with specific intelligent methods according to all (or most of) integration categories shown in Figure 1. For instance, combination of fuzzy RBR with CBR can follow the different coupling models concerning integration of RBR with CBR.
- In several application domains shown in Table 1, integrations of CBR with specific intelligent methods have not been applied.
- Implementation of (non-embedded) CBR combinations in other application domains besides the ones shown in Tables 1 and 2.
- Implementation of tightly coupled conditional sequence approaches in which the CBR component is invoked before the other component(s).

3. Representative Systems

In the following, some representative systems involving integration of CBR with other intelligent method(s) are presented in some detail, to give a better insight of the corresponding categories of the categorization scheme described in the previous section.

3.1 Sequential Processing Approaches

We present systems belonging to the sequential processing coupling category in two sections. One involves loosely coupled sequence approaches and the other one tightly coupled sequence approaches.

3.1.1 Loosely coupled sequence

The loosely coupled sequence approaches presented in this section come from [70], [16], [24], [8] and [48]. In all these approaches, except [48], the CBR component is invoked after the other component.

In [70] a general integrated approach for the classification task is presented. In this approach rules represent standard situations and cases represent exceptions or non-standard situations. The contents of the knowledge base are produced from an initial case base whose cases are split into two types: standard cases and exception cases. Standard cases are used to induce the rules of the knowledge base. The CBR module works with the exception cases. Splitting the initial case base is performed using heuristic approaches. For an input case, the inference process first checks if the rules can produce a conclusion. If they do, inference ends, otherwise CBR is employed. An advantage of the approach, as demonstrated by various experiments, is the good explanation ability stemming from the high level of comprehensibility of the rules. This is due to the fact that the rules induced from the standard cases are closer to expert rules than the rules produced from the whole dataset of cases (standard and exceptional). However, as is shown in [70], the splitting policy of the initial case base plays an important role in the accuracy and comprehensibility levels of the approach.

ECLAS [16] is a loan authorization system. The knowledge regarding the domain is discerned into two types: (a) objective, which is logical, explicit and rational and (b) subjective, which is implicit, uncertain and imprecise. RBR corresponds to the objective knowledge, whereas the subjective knowledge corresponds to CBR. During reasoning, the rule-based module is first invoked to process the input case (i.e., a loan application). If the rule-based module rejects the application, inference stops. Otherwise, if it approves it, the CBR module is invoked for further examination of the application so that the final decision on acceptance or rejection will be made. In ECLAS, the rule-based module filters several input cases that are rejected thus reducing the case match load of the case-based module.

In [24] a service-oriented event correlation approach is presented. Service fault management is important issue for service providers as it affects the quality of services delivered to customers, revenue (i.e. customer satisfaction) and costs concerning fault management itself and service level agreement violations. The approach performs automated event correlation by modeling services, resources and faults. Rules involve event, condition and action statements. The RBR component is first invoked and if it fails the CBR component is invoked. Advantages of the specific approach involve maintainability, modeling, robustness (i.e. ability to reach conclusions even when the knowledge base is inaccurate and ability to update knowledge base after a failed diagnosis) and time-performance.

In [8] an agent-based approach to manage emergency fires inside large oil storage and production plants is presented. Management involves fire-proof resource optimization and dangerous product evacuation. The approach concerns three different types of agents: a simulation agent to simulate physical/chemical phenomena and their consequences, a genetic agent to produce optimal management solutions and a CBR agent to adapt stored cases to the current scenario. Emergency process time is short (i.e. some minutes). The genetic agent is first invoked and if it is not able to produce a solution within specific time limits, the CBR agent is then invoked.

In [48] a medical system for the care of Alzheimer's disease patients is presented. The system provides decision support for neuroleptic drug prescription to patients with behavioral problems. The case-based module is invoked to determine if a neuroleptic drug should be prescribed to a patient and, if this is so, the rule-based module is invoked to select one of five such drugs. Such a system may improve the quality of life for Alzheimer's disease patients.

3.1.2 Tightly coupled sequence

The tightly coupled sequence approaches presented in this section come from [45], [71], [78], [20], [75], [43], [44], [15], [13], [35], [55], [65] and [41]. Table 3 depicts the tightly coupled sequence approaches involving compulsory sequence and the ones involving conditional sequence.

Table 3. Representative tightly coupled sequence approaches.

Compulsory Sequence	[45], [71], [78], [20] [75], [43], [44], [15]
Conditional Sequence	[13], [35], [55], [65], [41]

CAMPER [45] is a nutritional menu planner built by combining the best features of two independent menu planners, a case-based and a rule-based, namely CAMP and PRISM. Nutritional menu planning is a difficult task, because there are many numeric constraints some of which conflict with others, menus can be evaluated only if they are entirely constructed and common sense must be employed for combinations of foods that match or do not match. CAMP and PRISM were evaluated and compared, in order to locate their deficiencies and strengths. This analysis (resembling the standalone model) guided the construction of CAMPER. The CBR component constructs menus that are acceptable, since they satisfy multiple nutrition constraints. However, the rule-based component can enhance the proposed menu with new possibilities, by employing common sense and performing 'what if' analysis. Menus enhanced by rules are inserted into the case base, thus improving the effectiveness of the case-based module. CBR in CAMPER always produces an output that is subsequently improved by the invocation of rules (unless the menu produced by CAMP is deemed quite satisfactory). As in GYMEL [66], a significant reason for the usefulness of the combination is the difficulty in the acquisition of cases.

CORMS AI [71] is a real-time monitoring system assisting National Ocean Service watch standing personnel in its monitoring duties seven days per week. The system also includes a database to collect real-time sensor data. Based on the real-time data, the system invokes the rule-based module to identify problems and then the case-based module to recognize the source of each problem and to suggest appropriate remedial actions. CORMS AI has proven to be effective and robust during its operation decreasing the amount of time required by the personnel to identify and handle problems. It is estimated that the financial gain for the US government due to the operation of CORMS AI will be over one million dollars per year.

HIDES [78] is a system for diagnosing crop injury from herbicides. Although several intelligent systems have been developed in the weed science domain, no such system assisted in herbicide injury diagnosis. Herbicide diagnosis is a domain that is understood reasonably well but not perfectly and therefore integration of RBR and CBR offers benefits. Diagnosis is based on nine critical inputs. RBR is first invoked to identify suspect herbicide families, suspect herbicide(s) for causing the observed injury and to determine possible sources of the suspect herbicide(s). RBR also identifies the. These results are passed to CBR to propose a probable cause of injury (e.g. improper soil condition or herbicide carryover). The system can be used as an educational tool for both traditional classroom and extension classroom.

ScheduleCoach [20] is a system used to critique construction schedules. Due to the increasing complexity and scale of construction projects, construction schedules frequently contain errors that can be difficult to find. ScheduleCoach uses critique rules representing experts' critiquing principles to identify potential errors in a construction schedule. Cases represent previous successful projects. Some rules contain predetermined suggestions for the revision of objects violating schedule principles. Fired rules not containing such predetermined suggestions cause the invocation of the CBR module to determine appropriate revisions for the violating objects.

IDS [75] is a system used to improve aircraft fleet maintenance. It locates possible faults providing their complete description, the corresponding symptoms and the remedial actions. The system includes multiple rule bases performing different diagnostic actions. Rules take as inputs (real-time or offline) messages generated by diagnostic routines and locate faults providing their complete description as well as the corresponding symptoms. The case-based module is then invoked to find cases with similar symptoms and suggest appropriate remedial actions.

In [43], an approach combining CBR with fuzzy RBR is presented for risk prediction in environmental impact assessment. Environmental impact assessment concerns analyzing effects regarding development proposals before major decisions are taken and commitments are made. CBR is used to store past cases involving environmental impact statements and environmental impact assessment reports. Fuzzy RBR models expert knowledge concerning qualitative risk prediction. The linguistic terms used in fuzzy RBR provide naturalness and expressiveness in risk assessment. CBR is first invoked to retrieve similar past cases. Afterwards, fuzzy RBR is invoked taking as input the retrieved cases and performs qualitative risk prediction.

In [44] an approach combining CBR with a fuzzy belief network is described. The application domain concerns situation and threat assessments of ground battlespaces. Situation assessment infers relevant information about forces of concern in a military situation. It is a prerequisite to threat assessment which analyzes enemy intentions and capabilities. Situation assessment is performed by CBR and threat assessment by the fuzzy belief network. Four systems are invoked sequentially with the results of each system passed on to the next one in the sequence: three CBR systems and lastly a fuzzy belief network. All CBR systems take as input clustered features of detected ground target(s) in a specific area of the battlespace. The respective output produced by each CBR involve unit type, unit size and unit purpose. These three outputs are given as input to the fuzzy belief network.

HACM [15] concerns a conditional sequence approach to solve potential lawsuit problems caused by change orders in construction projects. The purpose is to avoid and resolve disputes before litigation occurs. HACM combines a back propagation neural network with CBR. The neural network is first invoked to determine whether there is likelihood for litigation concerning the given input case or not. If the neural network determines that there is no likelihood for litigation, reasoning ends. Otherwise the CBR module is invoked to retrieve similar past cases and displays warnings if degree of similarity exceeds 95%. The weights of the neural network are used to calculate similarity.

ELEM2-CBR [13] is a system integrating rules and cases to perform classification and numeric prediction. Rules are produced from cases using a rule induction method called ELEM2. However, in the reasoning process both rules and cases are used. Similarities between cases are determined in an innovative way by using relevance weighting. The induced rules are used to determine the weights of attribute-value pairs of the input case and cases in the case base are ranked according to their probability of relevance to the input case. Weights are calculated based on the relevant cases to the input case. For this purpose, the input case is matched against the induced rules. If matched rule(s) classify the input case to a single concept, cases belonging to that concept are considered relevant. If there are multiple matches, where rules classify the input case to different concepts, all cases belonging to those concepts are considered relevant. If no rule fires, rules partially matching the input case are ranked and the relevant cases are the ones belonging to the concept corresponding to the rules with the highest score. The numeric prediction task is mainly a case-based process using rules for weighting and relevance assessment. The classification task employs both RBR and CBR and returns the result of RBR if the input case is classified to a single concept or employs CBR, otherwise it uses the weighting relevance procedures described above. Experimental results comparing the accuracy of ELEM2-CBR with pure case-based methods or decision tree methods demonstrate its effectiveness.

URS-CBR [35] is a system used in Dutch Railways to classify images acquired from ultrasonic rail inspection. The amount of data (images) produced from ultrasonic inspection is huge and comes in a great variety making it necessary to minimize human intervention by performing automatic and reliable classifications. Efficiency, adaptability and maintenance were also prerequisites. Combination of rules with cases solved the problem. The system is made of two rule-based modules and a case-based module. For efficiency and maintenance reasons, cases are organized into a hierarchy of clusters and also the size of the rule bases is kept small. The first rule-based module precedes CBR. It takes as input the given image and tries to classify it. If it is successful in classifying the image, reasoning stops for that image. Otherwise, reasoning passes to the case-based module, which retrieves the most similar cases to the input case. To improve the efficiency of the retrieval process, intermediate conclusions reached by the first rule-based module are exploited for classification of the case to an appropriate cluster. For reliability reasons, the second rule-based module evaluates the retrieved cases in order to match them with the input case. Experiments were carried out comparing the hybrid system with pure rule-based and case-based classifiers. The results for the hybrid system showed an improvement in classification accuracy compared to both other systems. Its inference efficiency was worse than the pure rule-based system but better than the pure case-based system.

The system described in [55] is used for automated sleep stage scoring. The reason for using a hybrid system in that domain is the fact that human experts make decisions based on the combination of rules and experience. Rule-based knowledge is usually incomplete. The system consists of a signal processing unit, a rule-based and a case-based scoring unit. Rules are used to deal with usual situations and cases deal with details and exceptions to the rules. The rule-based module uses a simplified version of the certainty factor called the reliability value. In each reasoning phase, the rule-based scoring unit is first called. If the reliability value of the reached scoring conclusion exceeds a predefined threshold, the scoring process ends without invoking the case-based unit. If there are conflicts in applying rules or if no rule fires or if the reliability value of the reached conclusion is less than the threshold, the reasoning results of the rule-based unit are passed to the case-based unit that is invoked to make the decision. Cases include attributes regarding the applied rules and the conclusions of the rule-based reasoning process. These attributes play a role in similarity assessment. Experimental results showed an improvement in the accuracy of the hybrid system compared to pure rule-based or case-based systems.

In [65], a medical system for oncology is presented. Such a system can be used in hospital units to automate the process of checking whether a patient's case complies with appropriate guidelines or not. If it does not comply with guidelines, similar patients' cases will be exploited by experts to reach therapeutic decisions. Rules represent guidelines. A common restricted vocabulary is used for guidelines and cases. Key medical terms in both cases and guidelines are used to select the appropriate guideline with which to compare the case at hand. If the new case does not comply with the selected guideline, the results of RBR are used to determine similar cases to the input case. More specifically, the last guideline step with which the case complies is used to search for similar cases. The system is designed to be a data warehouse.

In [41] an approach for internal audits in banks is presented. Such an approach reduces risks, enables quick decision making for financial incidents and assists in upholding regulations, soundness and integrity of the financial system. Internal audits in banks usually involve time-consuming and tedious paperwork to examine numerous transactions as automatic audit systems are unusual. The approach belongs to conditional sequence subcategory. In the presented approach, RBR is invoked first to automatically detect suspicious transactions for which further actions are necessary. If such transactions are detected, the CBR component is invoked to scrutinize these transactions and provide punishment levels for involved employees. Rules formalize regulations and guidelines that should be upheld by employees. CBR works better than RBR in determining and recommending punishments since judgment is based on intuition and experience.

3.2. Co-processing Approaches

Presentation of representative co-processing approaches is organized in two sections. One section involves cooperation oriented approaches and the other one reconciliation oriented approaches.

3.2.1 Cooperation oriented

We present both types of representative cooperation oriented approaches: approaches employing explicit reasoning control and approaches employing implicit reasoning control. Table 4 depicts the co-processing employing explicit reasoning control and the ones employing implicit reasoning control.

Table 4. Representative cooperation oriented co-processing approaches.

Explicit Reasoning Control	[64], [51]
Implicit Reasoning Control	[9], [10], [11], [66], [34], [12], [31]

3.2.1.1 Explicit reasoning control

The presented cooperation oriented approaches employing explicit reasoning control are [64], [51].

CABARET [64] is an approach performing interpretation tasks in a legal reasoning domain (i.e. income tax law). CABARET consists of two co-reasoners, a rule-based and a case-based (having an equivalent status), a rule-based and a case-based monitor, a controller and a task agenda. The progress of each co-reasoner is observed by its associated monitor. The observations are described in a language understandable by the controller. The controller observes the operation of the whole system and each co-reasoner separately and decides how they will proceed in the reasoning process as a whole and individually. For this purpose, the controller uses a set of domain-independent heuristic rules encoded in the same language as the monitors' observations. Based on those heuristic rules, the controller adds, deletes or orders tasks for each co-reasoner on the agenda. The posted tasks enable the dynamic interleaving of the RBR and CBR processes. CABARET was reimplemented as a blackboard system.

The approach described in [51] integrates rules and cases in an innovative way. The approach has been applied to a medical domain, more specifically to diabetic patient management. The rule base of the system contains different classes of rules corresponding to different steps in the reasoning process. The innovative aspect is the ability to dynamically adapt rules belonging to specific classes in order to improve handling the new situation. Refinement of the rules is performed with the use of cases and involves certain parameters of the rules, which are too general to deal with the specific situation. Such parameters, for instance, are numeric thresholds appearing in conditions. The integration of RBR and CBR is controlled by a supervisor module that contains integration meta-rules. The integration makes the system more effective in detecting the patient's problems and providing enhanced prescriptions, thus reducing the time required to resolve the patient's problems.

3.2.1.2 Implicit reasoning control

The presented cooperation oriented approaches employing implicit reasoning control come from [9], [10], [11], [66], [34], [12] and [31].

CARE-PARTNER [9] proposes a framework for the close combination of the different knowledge base entities. Rules and cases are described using the same knowledge representation language. In this way, during inference, the knowledge base can be searched in parallel for applicable rules and cases enabling the reuse of all knowledge base entities. Pattern matching and case-based retrieval is performed in parallel and the conflict set may simultaneously contain rules as well as cases. Conflict resolution is based on two criteria: similarity to the input case and type of entities. Firstly, the conflict set entities are ranked according to their similarity degree to the input case and the most similar one is chosen. If there are two or more entities having the maximum similarity degree to the input case, a priority order is used giving preference to rules and then to cases. Therefore, the reasoning cycle tightly integrates the different knowledge base entities. This approach has been applied to a medical domain, more specifically to post-transplant patient care. A Web-based system has been developed for this purpose.

GREBE [10], [11] is an approach used in a legal reasoning domain generating arguments for specific point of views. GREBE uses a complex structured case representation scheme. More specifically, a semantic network representation is used to configure relations between case entities. Subgraphs of the graphical case representation relate facts relevant to a court decision concerning the satisfaction of the statutory predicate and those facts are the criterial facts of the case with respect to the predicate. In this way, GREBE's case representation is able to represent the relation between facts and results as determined by the court. The case-based reasoner possesses the mechanisms to efficiently handle the complex case representation. Its main actions are to retrieve the cases whose criterial facts most closely match those of the new case, to determine the best mapping from the criterial facts to the new case and to determine any facts that would improve the match if they were inferred. GREBE tries to solve its goals using both rules and cases providing to the user all the solutions it can find. An innovative reasoning aspect of GREBE is the ability to generate arguments created from parts of different cases and rules. The explanations produced from the synergy of the rule-based and case-based components are processed before shown to the user.

GYMEL [66] is a system for harmonizing melodies. Searching in the case base to find matching melodies proved to be a difficult problem and so each input problem was dealt as a set of simple problems. Each simple problem is to find a chord for a specific position based on information regarding this position and the previous chords in the sequence. If more than one chord is found for a position, backtracking is used to search all possibilities. For the solution of a simple problem, case-based reasoning is first invoked. Rules are invoked when the cases cannot produce a solution at a certain point during inference. The solution proposed by the rule-based module is passed to the case-based module that carries on inference. The approach is useful in application domains for which it is difficult to acquire an adequate set of cases and the CBR component needs to be backed up by a rule-based component expressing general knowledge. In such an approach, the invocation frequency of the rule-based component will be high at the early stages of the system's operation. Subsequently, however, it will decrease, as new cases will be incorporated into the case base.

In [34], an architecture concerning the analysis of event-based dynamic situations is described. Such a system could contribute to the understanding and awareness of complex scenarios such as homeland security threats and future battlespace engagements. The approach combines event correlation/management with situation awareness. More specifically, RBR is used for spatio-temporal event correlation and CBR for situation awareness. The rule-based and case-based modules act in a distributed fashion with each module dynamically invoking the other during inference.

SHYSTER-MYCIN [12] is a hybrid system used in the legal domain of copyright law. It combines SHYSTER, a case-based legal expert system, with MYCIN, a rule-based expert system (Buchanan and Shortliffe 1984). In this integration scheme, MYCIN was altered in a few aspects: its reporting was improved and no certainty factors were used due to the difficulty in defining them in this legal domain. Reasoning in SHYSTER-MYCIN focuses on the rules consulting cases, when an open textured term is met. However, there is no underlying control strategy for the invocation of SHYSTER and evaluation of its results. The system consults the user whether SHYSTER will be called when an open textured term is met or whether he/she can give an answer based on his/her knowledge. If SHYSTER is called, the user passes its reasoning results to MYCIN with the capability to make changes. Also, MYCIN and SHYSTER do not share facts and the user himself/herself has to pass data from one module to the other (the authors mention that a future version of the system will deal with this). Special care has been taken for testing SHYSTER-MYCIN. SHYSTER-MYCIN was tested against three criteria: validity, conciseness and correctness.

In [31] a cooperation-oriented approach in an intelligent Web-based e-learning system is presented. It provides personalized curriculum sequencing, a technology used in Intelligent Tutoring Systems, which involves selection, ordering and construction of the most appropriate teaching material and operations for a specific learner. This is very helpful for each learner because individual learning goals can be achieved more effectively. In contrast to other curriculum sequencing approaches, this approach simultaneously takes more parameters into consideration such as curriculum difficulty level, concept relation degrees of the curriculum, learner test responses in curriculum items, curriculum continuity of successive curriculums. Based on the aforementioned parameters, a genetic algorithm generates personalized curriculum sequencing. A genetic algorithm is useful due to the large search space. CBR performs summative assessment analysis. Summative assessment concerns a large portion of the course (e.g. two or more instruction units). CBR also provides capability to support corrective activities and formative assessment for an individual learner within a specific instruction unit.

3.2.2 Reconciliation oriented

The presented reconciliation oriented approaches come from [23], [40], [3], [5], [19], [58] and [72].

ANAPRON [23] involves combination of independent rules and cases in order to deal with the incompleteness and partial correctness of rules. Rules index cases, supporting them or contradicting them (exception cases), facilitating their retrieval. Exception cases fill the gaps of symbolic rules in representing domain knowledge. Therefore this approach results primarily in accuracy improvement of the rule-based component and secondarily in efficiency improvement of the case-based component. RBR competes with CBR in drawing conclusions. Inference focuses mainly on the symbolic rules, calling CBR only when necessary. The similarity metric of the case-based module returns a similarity score (i.e., a numerical rating of the similarity) and an analogical rule defining implicitly the analogy. During reasoning, firing of a rule is suspended when a sufficient number of its conditions are satisfied and its exception cases are checked. If an exception case is found having compelling analogy with the input case, the rule is not allowed to fire and the conclusion proposed by the retrieved exception case is considered valid instead. Decision regarding compellingness is based on a combination of criteria. More specifically, the similarity score between the exception and the input case, the accuracy and the significance of the analogical rule must exceed predefined thresholds. The accuracy and the significance of the analogical rule is estimated by taking into account both supporting and exception cases indexed by the symbolic rule. ANAPRON has been used in an application field defining the pronunciation of American surnames. This is a difficult task, due to the diverse national origins of the surnames. Experimental results have demonstrated the effectiveness of the combination, since ANAPRON approximates the performance of commercial systems in the domain. CCAR [40] handles inference as ANAPRON with the difference that only exception cases (and not cases supporting rules) are stored in the case base in order to improve case searching efficiency.

CoRCase [3] uses CBR to improve the real-time problem-solving capabilities of RBR used in a classification task. It can be thought of as an extension to the approach used in ANAPRON. Different types of indices are employed for the cases according to all the roles they play in rule-based problem solving. A case that has been solved successfully by the system (i.e., the system outcome is confirmed by the expert) is indexed as true positive by the solution found and as true negative by each rejected solution during problem solving. An erroneously solved case is indexed as false positive by the rule it satisfies and as false negative with respect to the category representing its real solution that has been rejected during problem solving. Indices are used after the invocation of the RBR component to analyze the case at hand and determine similar stored cases. Reconciliation is used to deal with two situations: when there is an indication that the expert will reject the rule-based solution (due to past experience) and when RBR cannot produce a solution to the problem at hand. The conclusion produced by the system corresponds to the best-matched case. The retrieval process takes into account rules as well due to the fact that rules themselves are considered as generalized cases.

In [5] among others, a combination of a neural network and CBR according to the reconciliation coupling approach is presented for accuracy improvement. The case base consists of the neural network training examples and each case is indexed by its real solution and by its neural network solution. The approach can employ any type of neural network performing classification.

MARS [19] is a hybrid system used in the financial domain of mergers and acquisitions. The system achieves a seamless combination of RBR and CBR within one architectural framework, that of RBR. The system uses possibilistic reasoning to represent uncertainty and imprecision underlying the reasoning process. Rules are associated with a sufficiency measure (indicating the strength with which the antecedent implies the consequent), a necessity measure (indicating the degree to which a failed antecedent implies a negated consequent) and a context, which represents the set of preconditions determining the rule's applicability to a given situation. Cases are implemented as rule templates. To achieve this conversion, information such as sufficiency, necessity and context is needed for each case. CBR is activated at specific situations determined by the system designer. Rules and cases are considered as separate proof paths to a conclusion, making proportional contribution or disconfirmation of the conclusion.

In [58] an approach for lung disease diagnosis is presented. The approach combines a CBR component using fuzzy terms in case representation with a fuzzy RBR component. The case base consists of patient records whereas rules encompass doctor experience. Both modules are invoked in parallel and a type of numeric reconciliation is performed: the similarity value of the most relevant case and the conclusion degree of the fired rule are averaged to produce a more accurate and realistic conclusion degree. Both combined approaches contribute to the diagnosis with the same weight in case they diagnose the same disease. In case they diagnose different diseases, the combination of the components' conclusions cannot be done.

In [72] an approach combining fuzzy CBR with fuzzy RBR is presented. The application domain involves treatment planning for adolescent early intervention of mental healthcare. The specific domain is crucial and complex. Rules represent experience of social service professionals whereas cases client records. The RBR and CBR components are invoked in parallel. The corresponding results are combined according to specific formulas.

3.3 Embedded Processing

As already mentioned, embedded processing approaches give pre-eminence to CBR or pre-eminence to other intelligent method (Figure 2).

Embedded processing approaches giving pre-eminence to CBR involve CBR systems employing one or more modules of other representation methods to perform their internal (offline and online) tasks. Typical such CBR tasks involve retrieval and adaptation. Retrieval concerns several procedures such as situation assessment, employing preferences, exclusion criteria and heuristic procedures in case selection [38]. Adaptation is a time-consuming and complex task often requiring domain-specific knowledge [38], [50]. A single intelligent method (e.g. genetic algorithm or neural network) may be employed in different CBR tasks.

Indicative internal CBR tasks that can be performed by other intelligent methods are the following:

- *Initial case base construction.* In domains with insufficient amount of available cases, intelligent methods such as RBR [54], fuzzy methods [69] and genetic algorithms [68].
- *Maintenance of case base.* Maintenance tasks play a significant role in time-performance and accuracy of a CBR [18], [74]. For such tasks, intelligent methods such as rules, genetic algorithms (Ahn and Kim 2009), neural networks [49] and fuzzy methods [67] may be employed.
- *Case representation.* For case representation, methods such as genetic algorithms [36], neural networks [49] and fuzzy methods [72] may be used.
- *Case retrieval.* To retrieve useful past cases, various intelligent methods may be used such as RBR [38], fuzzy methods [42], neural networks [79] and genetic algorithms [76].
- *Case adaptation.* To perform case adaptation, methods such as domain-specific and domain-independent rules ([38], [50], [37]), neural networks [22] and genetic algorithms [32].

An embedded approach of this type is DIAL [37]. DIAL is a system developed for disaster response planning and effectively deals with a main problem of case-based systems that is, the acquisition of case adaptation knowledge. The innovative idea of this system is to acquire adaptation knowledge during its operation in the form of cases. Another benefit of case-based adaptation knowledge is its adaptability, in contrast to rule-based adaptation knowledge, enabling the generation of more effective plans. Furthermore, similarity measures are dynamically adapted based on the acquired case-based adaptation knowledge. Multiple cooperating rule-based and case-based components are incorporated into the case-based planner in order to perform the adaptation and similarity tasks. Rule-based adaptation knowledge consists of general abstract rules and rule-based similarity knowledge corresponds to predefined domain-specific criteria. The system tries to perform each task by calling the case-based component falling back on the rule-based component in case of failure. The advantages of the system are the improved inference efficiency and the generation of better plans compared to a conventional case-based system.

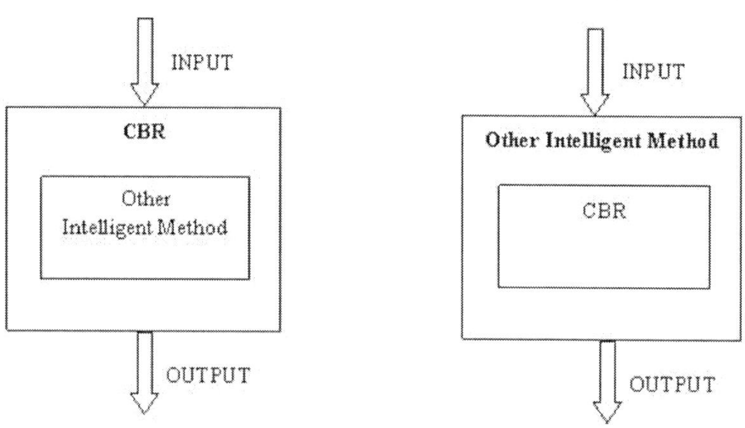

Figure 2. Embedded Processing Model: pre-eminence to CBR (left), pre-eminence to other method (right).

Embedded approaches giving pre-eminence to the other intelligent method are less usual however they can be interesting as far as hybridism is concerned. Such implemented approaches mainly involve use of CBR to enhance genetic algorithms.

CBR may improve genetic algorithms in the following ways:

- *Tuning genetic algorithm parameters* such as population size, crossover rate and mutation rate [56]. The values of these parameters play an important role in the performance of a genetic algorithm.
- *Exploiting stored cases to enhance genetic process*. Phases of the genetic process may be appropriately stored as cases and used subsequently as part of the chromosome population [57]. Such approaches may reduce convergence time and improve accuracy of genetic algorithms.
- *Exploiting stored cases for provision of explanations and knowledge extraction* [57]. Genetic algorithms do not provide explanations for reached solutions. Deriving knowledge regarding the genetic algorithm process may be useful to the implementation of future genetic algorithms. Stored cases corresponding to phases of the genetic process may contribute in handling such issues.

Implementation of approaches embedding CBR within other intelligent methods besides genetic algorithms could be an interesting future direction. Such an approach could exploit accumulated experience to improve internal tasks of intelligent methods (e.g. by learning from successes and failures).

4. Combination of CBR with Neurules

In the following, we describe an approach combining CBR with neurules, a type of hybrid rules integrating symbolic rules with a neural component (i.e. adaline unit) in a uniform/seamless way [59], [28]. The integrated approach belongs to the reconciliation oriented co-processing category.

Neuro-symbolic approaches combine neural and symbolic approaches. A large part of such approaches combine symbolic rules with neural networks. Such combinations have produced effective representation formalisms due to the complementary advantages/ disadvantages of the combined approaches [28].

Rules offer a number of advantages for knowledge representation such as, naturalness, modularity, interactive inference mechanisms enabling inference tracing by humans and explanation mechanisms providing explanations regarding inference process. Naturalness facilitates comprehension of knowledge represented by rules whereas modularity refers to the fact that each rule is an autonomous unit. However, rules exhibit certain drawbacks such as difficulty in knowledge acquisition from experts, inability to exploit experience in inference, inference efficiency problems in very large rule bases and inability to draw conclusions in case of missing values in input data or in case of unexpected input values. Neural networks exhibit advantages such as knowledge acquisition from training examples, representation of complex knowledge, efficiency in producing outputs and generalization capabilities. On the other hand, neural networks lack the naturalness and modularity of symbolic rules making it difficult to comprehend their encompassed knowledge, do not provide interactive inference mechanisms and do not provide explanations for reached output.

Most neuro-symbolic approaches resulting into a uniform/seamless combination of the symbolic and neural components give pre-eminence to the neural component. More specifically, the neural component is the main one in which symbolic knowledge is incorporated in or mapped to. In this way, most neuro-symbolic approaches lack the advantages of symbolic rules. In contrast to such approaches, neurules give pre-eminence to the symbolic component retaining naturalness and modularity of symbolic rules and also providing interactive inference mechanisms and explanation facilities [25], [26], [27], [30]. Neurule-based reasoning is more efficient than symbolic RBR [25]. Also in contrast to symbolic rules, conclusions can be reached from neurules even if some of the conditions are unknown. Finally, neurules generalize quite well [30].

4.1 Syntax and Semantics

Neurules are a kind of integrated rules. The form of a neurule is depicted in Figure3a. Each condition C_i is assigned a number sf_i, called its *significance factor*. Moreover, each rule itself is assigned a number sf_0, called its *bias factor*. Internally, each neurule is considered as an adaline unit (Figure3b). The *inputs* C_i (i=1,...,n) of the unit are the conditions of the rule. The weights of the unit are the significance factors of the neurule and its bias is the bias factor of the neurule. Each input takes a value from the following set of discrete values: [1 (true), -1 (false), 0 (unknown)].

The *output* D, which represents the conclusion (decision) of the rule, is calculated via the standard formulas:

$$D = f(a), \quad a = sf_0 + \sum_{i=1}^{n} sf_i C_i \qquad (1)$$

$$f(a) = \begin{cases} 1 & \text{if } a \geq 0 \\ -1 & \text{otherwise} \end{cases} \qquad (2)$$

where ***a*** is the *activation value* and *f(x)* the *activation function*, which is a threshold function. Hence, the output can take one of two values ('-1', '1') representing failure and success of the rule respectively. The significance factor of a condition represents the significance (weight) of the condition in drawing the conclusion.

The general syntax of a neurule (in a BNF notation, where '< >' denotes non-terminal symbols) is:

<rule>::= (<bias-factor>) if <conditions> then <conclusion>
<conditions>::= <condition> | <condition>,<conditions>

<condition>::= <variable> <l-predicate> <value> (<significance-factor>)
<conclusion>::= <variable> <r-predicate> <value> .

where <variable> denotes a *variable*, that is a symbol representing a concept in the domain, e.g. 'sex', 'pain' etc in a medical domain, and <l-predicate> denotes a symbolic or a numeric predicate. The symbolic predicates are {is, isnot}, whereas the numeric predicates are {<, >, =}. <r-predicate> can only be a symbolic predicate. <value> denotes a value; it can be a symbol (e.g. "male", "night-pain") or a number (e.g "5"). <bias-factor> and <significance-factor> are (real) numbers.

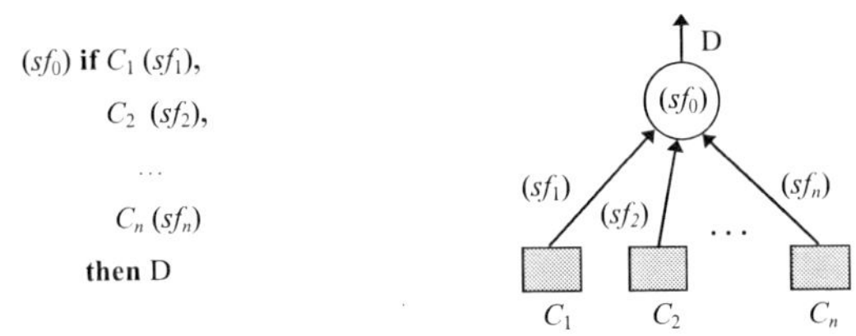

Figure. 3. (a). Form of a neurule, (b) a neurule as an adaline unit.

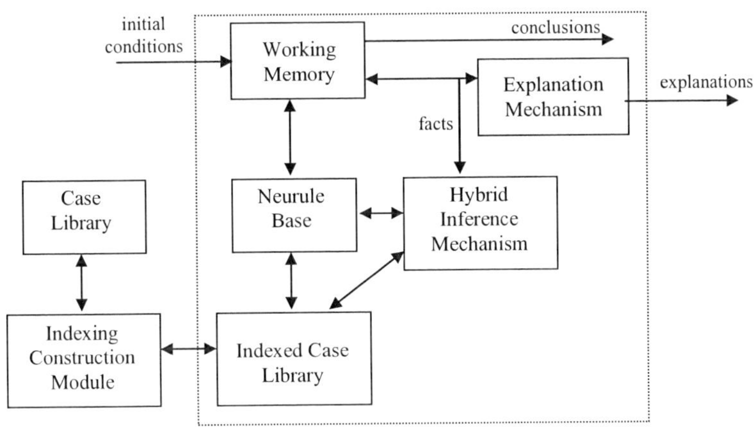

Figure 4. Architecture of a system integrating CBR with neurules.

A variable in a condition can be either an input variable or an intermediate variable or even an output variable, whereas a variable in a conclusion can be either an intermediate or an output variable. An input variable takes values from the user (input data), whereas intermediate or output variables take values through inference since they represent intermediate and final conclusions respectively.

Neurules are constructed either from empirical data (i.e. training examples) [26] or from symbolic rules [25] thus exploiting existing symbolic rule bases. In either creation process, an adaline unit is initially assigned to each of the intermediate and final conclusions. Each unit is individually trained via the Least Mean Square (LMS) algorithm. If the patterns in the training set of a neurule form a non-separable set, special techniques are used. In that case, more than one neurule having the same conclusion are produced. When neurules are produced from symbolic rules, each neurule usually corresponds to (or merges) a set of symbolic rules called its merger set [25]. Therefore, the size of the neurule base is reduced compared to the size of the corresponding symbolic rule base.

The neurule-based inference engine gives pre-eminence to symbolic reasoning, based on a backward chaining strategy [25]. Conclusions are reached based on the values of the condition variables and the weighted sums of the conditions. A neurule fires if the output of the corresponding adaline unit is computed to be '1' after evaluation of its conditions. A neurule is said to be 'blocked' if the output of the corresponding adaline unit is computed to be '-1' after evaluation of its conditions. To facilitate inference, conditions of neurules are organized according to the descending order of their significance factors. When a neurule is examined during inference, certain heuristics are applied to avoid evaluation of all its conditions [25].

4.2 Indexing and Hybrid Inference

Figure 4 depicts the architecture of a system integrating neurule-based and case-based reasoning. The run-time system (in the dashed shape) consists of the following modules: the *working memory*, the *hybrid inference mechanism*, the *explanation mechanism*, the *neurule base* and the *indexed case library*. The neurule base contains neurules. Neurules index cases representing their exceptions. The *indexing construction module* implements the process of acquiring an indexing scheme. The indexing process takes as input the following two types of knowledge:

- *Available neurules and cases*. The indexing scheme for this type of knowledge is acquired by performing neurule-based reasoning for the neurules based on the attribute values of each case. Whenever a neurule fires and the value of the conclusion variable does not match the corresponding attribute value of a case, the case is marked as an exception to this neurule.
- *Available symbolic rules and exception cases*. This type of knowledge concerns an available formalism of symbolic rules and indexed exception cases as the one presented in [23]. The indexing scheme for this type of knowledge is acquired by converting symbolic rules to neurules. The produced neurules are associated with the exception cases of the symbolic rules belonging to their merger sets.

The hybrid inference mechanism combines neurule-based and case-based reasoning by considering facts contained in the working memory, neurules in the neurule base and cases in the indexed case library. More specifically, the hybrid inference process focuses on neurules (i.e. neurule-based reasoning). If an adequate number of the conditions of a neurule are fulfilled so that it can fire, firing of the neurule is suspended and CBR is performed for its indexed exception cases. CBR results are evaluated as in [23] to assess whether the neurule will fire or whether the conclusion proposed by the exception case will be considered valid.

Results have shown the effectiveness of the approach [59], [28]. Cases can be used to fill neurule gaps in representing domain knowledge. Therefore, integration of CBR with neurules primarily improves accuracy of the overall system. Integration results in accuracy improvement regardless the source knowledge type of neurules (i.e. symbolic rules or empirical data) [59], [28], [62]. Furthermore, if the knowledge source of the integrated system concerns an available formalism of symbolic rules and indexed exception cases as the one presented in [23], inference is performed more efficiently [59], [28]. Neurules are a type of hybrid rules and thus one could compare our approach with approaches combining RBR with CBR. The approach combining neurules with CBR offers advantages such as more efficient inferences and drawing of conclusions even if certain input values are unknown.

Due to the fact that neurules seamlessly integrate symbolic rules with a neural component, the specific approach integrates three intelligent methods: symbolic rules, neural networks and CBR. Few (non-embedded) CBR integrations involve more than two combined approaches. Furthermore, the approach offers advantages such as naturalness, modularity, provision of explanations for drawn conclusions and exploitation of different knowledge sources.

It should be mentioned that combinations of neuro-symbolic approaches with CBR are quite rare. Such combinations could be an interesting research direction as they could exploit different types of knowledge sources such as symbolic domain knowledge (usually in the form of rules), training examples and case-based knowledge.

Another approach integrating a neuro-symbolic method with CBR is presented in [4]. Integration follows the reconciliation oriented co-processing approach. The specific neuro-symbolic method concerns a type of knowledge-based neural network. Knowledge-based neural networks are neural networks to which initial symbolic domain knowledge is mapped. The specific approach lacks advantages of our approach such as naturalness, modularity and ability to provide explanations.

An interesting future direction in the integration of CBR with neurules involves use of different types of case indices besides 'exception' indices. Initial results towards this direction are promising [62]. An additional future direction involves maintenance of the integrated representation scheme in case of updates in the neurule source knowledge (i.e. symbolic rule base or training examples). In [60], mechanisms for efficiently updating a neurule base due to changes to its symbolic source knowledge (i.e. symbolic rule base) are presented. Changes to the symbolic rule base involve insertion of a new symbolic rule or removal of an old rule. The presented mechanisms efficiently update the neurule base due to such changes to the source knowledge by storing information related to the neurule construction process to a tree, called the splitting tree. These update mechanisms should be extended and revised to accommodate a formalism integrating CBR with neurules.

CONCLUSIONS

In this chapter, we discuss issues involving integrations of CBR with other intelligent methods. Several such approaches have been developed. We categorize CBR integrations, briefly present representative systems applied in various domains and outline directions for future work. We also discuss issues involving combination of CBR with neurules, a neuro-symbolic method retaining advantages of symbolic rules.

CBR integrations concern an important area for AI researchers. Working processes in most fields have been automated with the use of various information systems. Such systems record case data in electronic form. Therefore, an abundant amount of cases is available in several domains. Such data can be exploited in development of integrated intelligent systems when deemed necessary facilitating knowledge acquisition. Research fields involving other combinations (e.g. neuro-symbolic or neuro-fuzzy methods) have been extensively explored. Various types/categories of such combinations have been implemented. This remains to be done for CBR integrations.

REFERENCES

[1] Aamodt, A. & Plaza, E. (1994). Case-Based Reasoning: Foundational Issues, Methodological Variations, and System Approaches. *AI Communications*, *7*, 39-59.

[2] Abraham, A. (2003). Intelligent Systems: Architectures and Perspectives. In J., Kacprzyk, A. Abraham, & L. C. Jain, (Eds.), Recent Advances in Intelligent Paradigms and Applications: *Studies in Fuzziness and Soft Computing*, (1-35). Heidelberg, New York: Physica-Verlag.

[3] Agre, G. (1995). KBS Maintenance as Learning Two-Tiered Domain Representation. In M. Veloso, & A. Aamodt, (Eds.), Case-Based Reasoning Research and Development: *Lecture Notes in Artificial Intelligence*, (*vol. 1010*, 108-120). Berlin, Heidelberg, New York: Springer-Verlag.

[4] Agre, G. & Koprinska, I. (1996). Case-Based Refinement of Knowledge-Based Neural Networks. In J., Albus, A. Meystel, & R. Quintero, (Eds.), Proceedings of the International Conference "Intelligent Systems: A Semiotic Perspective" (vol. II: *Applied Semiotics*, 221-226), Gaithersburg, MD: US Government Printing Office.

[5] Agre, G. (1998). A Two-Tiered Reasoning and Learning Architecture. *In Proceedings of the 1998 AAAI Spring Symposium on Multimodal Reasoning*, AAAI Technical Report SS-98-04 (124-129), Menlo Park, CA: AAAI Press.

[6] D. Aha, & J. J. Daniels, (Eds.) (1998). Case-Based Reasoning Integrations: Papers from the 1998 AAAI Workshop. *Technical Report*, WS-98-15. Menlo Park, CA: AAAI Press.

[7] Bader, S. & Hitzler, P. (2005). Dimensions of Neural-Symbolic Integration – a Structured Survey. In S., Artemov, H., Barringer, A. S., D'Avila Garcez, L. C. Lamb, & J. Woods, (Eds.), We Will Show Them: Essays in Honour of Dov Gabbay (volume 1, 167-194). *International Federation for Computational Logic*, College Publications.

[8] Balducelli, C. & D' Esposito, C. (2000). Genetic Agents in an EDSS System to Optimize Resources Management and Risk Object Evacuation. *Safety Science*, *35*, 59-73.

[9] Bichindaritz, I., Moinpour, C., Donaldson, G., Bush, N., Kansu, E. & Sullivan, K. M. (2003). Case-Based Reasoning for Medical Decision-Support in a Safety Critical Environment. In M., Dojat, E. Keravnou, & P. Barahona, (Eds.), Artificial Intelligence in Medicine Europe: *Lecture Notes in Artificial Intelligence*, (*vol. 2780*, 314-323). Berlin, Heidelberg, New York: Springer-Verlag.

[10] Branting, L. K. (1991). Building Explanations from Rules and Structured Cases. *International Journal of Man-Machine Studies*, *34*, 797-837.

[11] Branting, L. K. (2003). A Reduction-Graph Model of Precedent in Legal Analysis. *Artificial Intelligence*, *150*, 59-95.

[12] O'Callaghan, T. A., Popple, J. & Mc Creath, E. (2003). SHYSTER-MYCIN: A Hybrid Legal Expert System, *in Proceedings of the Ninth International Conference on AI in Law*, (103-104), New York, NY: ACM Press.

[13] Cercone, N., An, A. & Chan, C. (1999). Rule-Induction and Case-Based Reasoning: Hybrid Architectures Appear Advantageous. *IEEE Transactions on Knowledge and Data Engineering*, *11*, 164-174.

[14] Cheetam, W., Shiu, S. C. K. & Weber, R. O. (2006). Soft Case-Based Reasoning. *Knowledge Engineering Review*, *20*, 267-269.

[15] Chen, J. H. & Hsu, S. C. (2007). Hybrid ANN-CBR Model for Disputed Change Orders in Construction Projects. *Automation in Construction*, *17*, 56–64.

[16] Chen, H. P. & Wilkinson, L. J. (1998). *Case Match Reduction through the Integration of Rule-based and Case-Based Reasoning Procedures*. In [6] (33-38).

[17] Chi, R. T. H. & Kiang, M. Y. (1993). Reasoning by Coordination: an Integration of Case-Based and Rule-Based Reasoning Systems. *Knowledge-Based Systems*, *6*, 103-113.

[18] De Mantaras, R. L., McSherry, D., Bridge, D., Leake, D., Smyth, B., Craw, S., Faltings, B., Maher, M. L., Cox, M. T., Forbus, K., Keane, M., Aamodt, A. & Watson, I. (2005). Retrieval, Reuse, Revision, and Retention in Case-Based Reasoning. *The Knowledge Engineering Review*, *20*, 215-240.

[19] Dutta, S. & Bonissone, P. P. (1993). Integrating Case Based and Rule Based Reasoning. *International Journal of Approximate Reasoning*, *8*, 163-203.

[20] Dzeng, R. J. & Lee, H. Y. (2004). Critiquing Contractors' Scheduling by Integrating Rule-Based and Case-Based Reasoning. *Automation in Construction*, *13*, 665-678.

[21] Evans-Romaine, K. & Marling, C. (2003). Prescribing Exercise Regimens for Cardiac and Pulmonary Disease Patients with CBR. In Proceedings of Workshop on CBR in the Health Services, 5th *International Conference on Case-Based Reasoning*, (45-52).

[22] Fdez-Riverola, F. & Corchado, J. M. (2004). FSfRT: Forecasting System for Red Tides. *Applied Intelligence*, *21*, 251-264.

[23] Golding, A. R. & Rosenbloom, P. S. (1996). Improving Accuracy by Combining Rule-Based and Case-Based Reasoning. *Artificial Intelligence*, *87*, 215-254.

[24] Hanemann, A. & Marcu, P. (2008). Algorithm Design and Application of Service-Oriented Event Correlation. *In Proceedings of the 3rd IEEE/IFIP International Workshop on Business-driven IT Management*, (61-70). IEEE.

[25] Hatzilygeroudis, I. & Prentzas, J. (2000). Improving the Performance of Symbolic Rules. *International Journal on Artificial Intelligence Tools*, *9*, 113-130.

[26] Hatzilygeroudis, I. & Prentzas, J. (2001a). Constructing Modular Hybrid Rule Bases for Expert Systems. *International Journal on Artificial Intelligence Tools*, *10*, 87-105.

[27] Hatzilygeroudis, I. & Prentzas, J. (2001b). An Efficient Hybrid Rule-Based Inference Engine with Explanation Capability. In J. Kolen, & I. Russell, (Eds.), *Proceedings of the Fourteenth International Florida Artificial Intelligence Research Society Conference*, (227-231). Menlo Park, CA: AAAI Press.

[28] Hatzilygeroudis, I. & Prentzas, J. (2004a). Integrating (Rules, Neural Networks) and Cases for Knowledge Representation and Reasoning in Expert Systems. *Expert Systems with Applications*, *27*, 63-75.

[29] Hatzilygeroudis, I. & Prentzas, J. (2004b). Neuro-Symbolic Approaches for Knowledge Representation in Expert Systems. *International Journal of Hybrid Intelligent Systems*,

1, 111-126.

[30] Hatzilygeroudis, I. & Prentzas, J. (2010). Integrated Rule Based Learning and Inference. *IEEE Transactions on Knowledge and Data Engineering*, (in press).

[31] Huang, M. J., Huang, H. S. & Chen, M. Y. (2007). Constructing a Personalized E-learning System Based on Genetic Algorithm and Case-Based Reasoning Approach. *Expert Systems with Applications*, *33*, 551–564.

[32] Huang, B. W., Shih, M. L., Chiu, N. H., Hu, W. Y. & Chiu, C. (2009). Price Information Evaluation and Prediction for Broiler using Adapted Case-Based Reasoning Approach. *Expert Systems with Applications*, *36*, 1014-1019.

[33] Jacobo, V. H., Ortiz, A., Cerrud, Y. & Schouwenaars, R. (2007). Hybrid Expert System for the Failure Analysis of Mechanical Elements. *Engineering Failure Analysis*, *14*, 1435-1443.

[34] Jakobson, G., Buford, J. & Lewis, L. (2004). Towards an Architecture for Reasoning about Complex Event-Based Dynamic Situations. In Proceedings of the Third International Workshop on Distributed Event Based Systems, *in conjunction with the 26th International Conference on Software Engineering*, (62-67), IEE.

[35] Jarmulak, J., Kerckhoffs, E. J. H. & Van't Veen, P. P. (2001). Case-based Reasoning for Interpretation of Data from Non-Destructive Testing. *Engineering Applications of Artificial Intelligence*, *14*, 401-417.

[36] Kim, K. J. (2004). Toward Global Optimization of Case-Based Reasoning Systems for Financial Forecasting. *Applied Intelligence*, *21*, 239-249.

[37] Kinley, L. (2001). Learning to Improve Case Adaptation. PhD Thesis, *Computer Science Department*, Indiana University, USA.

[38] Kolodner, J. L. (1993). *Case-Based Reasoning*. San Mateo, CA: Morgan Kaufmann.

[39] D. B. Leake, (Ed.) (1996). Case-Based Reasoning: Experiences, Lessons, and Future Directions. *Menlo Park*, CA: AAAI Press.

[40] Lee, M. R. (2002). An Exception Handling of Rule-Based Reasoning Using Case-Based Reasoning. *Journal of Intelligent and Robotic Systems*, *35*, 327-338.

[41] Lee, G. H. (2008). Rule-Based and Case-Based Reasoning Approach for Internal Audit of Bank. *Knowledge-Based Systems*, *21*, 140-147.

[42] Li, S. T. & Ho, H. F. (2009). Predicting Financial Activity with Evolutionary Fuzzy Case-Based Reasoning, *Expert Systems with Applications*, *36*, 411-422.

[43] Liu, K. F. R. & Yu, C. W. (2009). Integrating Case-Based and Fuzzy Reasoning to Qualitatively Predict Risk in an Environmental Impact Assessment Review. *Environmental Modelling & Software*, *24*, 1241-1251.

[44] Looney, C. G. & Liang, L. R. (2003). Cognitive Situation and Threat Assessments of Ground Battlespaces. *Information Fusion*, *4*, 297-308.

[45] Marling, C., Petot, G. J. & Sterling, L. S. (1999). Integrating Case-Based and Rule-Based Reasoning to Meet Multiple Design Constraints. *Computational Intelligence*, *15*, 308-332.

[46] Marling, C., Rissland, E. & Aamodt, A. (2006). Integrations with Case-Based Reasoning. *Knowledge Engineering Review*, *20*, 241-245.

[47] Marling, C., Sqalli, M., Rissland, E., Munoz-Avila, H. & Aha, D. (2002). Case-Based Reasoning Integrations. *AI Magazine*, *23*, 69-86.

[48] Marling, C. & Whitehouse, P. (2001). Case-Based Reasoning in the Care of Alzheimer's Disease Patients. In D. Aha, & I. Watson, (Eds.), Case-Based Reasoning

Research and Development: *Lecture Notes in Computer Science*, (vol. 2080, 702-715). Berlin, Heidelberg, New York: Springer-Verlag.

[49] Medsker, L. R. (1998). Hybrid Intelligent Systems (Second Printing). Norwell, MA: Kluwer Academic Publishers.

[50] Mitra, R. & Basak, J. (2005). Methods of Case Adaptation: *a Survey. International Journal of Intelligent Systems*, *20*, 627-645.

[51] Montani, S. & Bellazzi, R. (2002). Supporting Decisions in Medical Applications: the Knowledge Management Perspective. *International Journal of Medical Informatics*, *68*, 79-90.

[52] Nauck, D., Klawonn, F. & Kruse, R. (1997). *Foundations of Neuro-Fuzzy Systems*. New York, NY: John Wiley & Sons.

[53] Pal, S. K. & Shiu, S. C. K. (2004). *Foundations of Soft Case-Based Reasoning*. New York, NY: John Wiley & Sons.

[54] Pandey, B. & Mishra, R. B. (2009). An Integrated Intelligent Computing Model for the Interpretation of EMG Based Neuromuscular Diseases. *Expert Systems with Applications*, *36*, 9201-9213.

[55] Park, H. J., Oh, J. S., Jeong, D. U. & Park, K. S. (2000). Automated Sleep Stage Scoring Using Hybrid Rule- and Case-Based Reasoning. *Computers and Biomedical Research*, *33*, 330-349.

[56] Pavon, R., Dvaz, F., Laza, R. & Luzon, V. (2009). Automatic Parameter Tuning with a Bayesian Case-Based Reasoning System. A Case of Study. *Expert Systems with Applications*, *36*, 3407-3420.

[57] Perez, E. I., Coello Coello, C. A. & Hernandez-Aguirre, A. (2005). Extraction and Reuse of Design Patterns from Genetic Algorithms using Case-Based Reasoning. *Soft Computing*, *9*, 44-53.

[58] Phuong, N. H., Prasad, N. R., Hung, D. H. & Drake, J. T. (2001). Approach to Combining Case Based Reasoning with Rule Based Reasoning for Lung Disease Diagnosis. *In Proceedings of the Joint Ninth International Fuzzy Systems Association World Congress and Twentieth International Conference of the North American Fuzzy Information Processing Society*, (vol. 2, 883-888), IEEE.

[59] Prentzas, J. & Hatzilygeroudis, I. (2002). Integrating Hybrid Rule-based with Case-Based Reasoning. In S. Craw, & A. D. Preece, (Eds.), Advances in Case-Based Reasoning, *Lecture Notes in Computer Science*, (vol. 2416, 336-349). Berlin, Heidelberg, New York: Springer-Verlag.

[60] Prentzas, J. & Hatzilygeroudis, I. (2005). Rule-based Update Methods for a Hybrid Rule Base. *Data and Knowledge Engineering*, *55*, 103-128.

[61] Prentzas, J. & Hatzilygeroudis, I. (2007). Categorizing Approaches Combining Rule-Based and Case-Based Reasoning. *Expert Systems*, *24*, 97-122.

[62] Prentzas, J., Hatzilygeroudis, I. & Michail, O. (2008). Improving the Accuracy of Neuro-Symbolic Rules with Case-Based Reasoning. In I., Hatzilygeroudis, C. Koutsojannis, & V. Palade, (Eds.), *Proceedings of the First International Workshop on Combinations of Intelligent Methods and Applications in conjunction with 18th European Conference on Artificial Intelligence*, (49-54), Patras, Greece.

[63] Prentzas, J. & Hatzilygeroudis, I. (2009). Combinations of Case-Based Reasoning with Other Intelligent Methods. *International Journal of Hybrid Intelligent Systems*, *6*, 189-209.

[64] Rissland, E. L. & Skalak, D. B. (1991). CABARET: Rule Interpretation in a Hybrid Architecture. *International Journal of Man-Machine Studies*, *34*, 839-887.

[65] Rossille, D., Laurent, J. F. & Burgun, A. (2005). Modeling a Decision-Support System for Oncology Using a Rule-Based and Case-Based Reasoning Methodologies. *International Journal of Medical Informatics*, *74*, 299-306.

[66] Sabater, J., Arcos, J. L. & Lopez De Mantaras, R. (1998). *Using Rules to Support Case-Based Reasoning for Harmonizing Melodies*, In [6] (147-151).

[67] Shiu, S. C. K., Li, Y. & Wang, X. Z. (2001). Using Fuzzy Integral to Model Case-Base Competence. In Proceedings of the Workshop on Soft Computing in Case-Based Reasoning, *Fourth International Conference in Case-Based Reasoning*, (206-212).

[68] Soh, L. K. & Tsatsoulis, C. (2001). Combining Genetic Algorithms and Case-Based Reasoning for Genetic Learning of a Casebase: A Conceptual Framework. *In Proceedings of the Genetic and Evolutionary Computation Conference*, (376-383). San Mateo, CA: Morgan Kaufmann Publishers.

[69] Sun, Z., Finnie, G. & Weber, K. (2004). Case Base Building with Similarity Relations. *Information Sciences*, *165*, 21-43.

[70] Surma, J. & Vanhoof, K. (1998). *An Empirical Study on Combining Instance-Based and Rule-Based Classifiers. In*, [6] (130-134).

[71] Vafaie, H. & Cecere, C. (2005). CORMS AI: Decision Support System for Monitoring US Maritime Environment. In N. Jacobstein, B. Porter, (Eds.), *Proceedings of the Seventeenth Innovative Applications of Artificial Intelligence Conference*, (1499-1506). Menlo Park, CA: AAAI Press.

[72] Wang, W. M., Cheung, C. F., Lee, W. B. & Kwok, S. K. (2007). Knowledge-Based Treatment Planning for Adolescent Early Intervention of Mental Healthcare: a Hybrid Case-Based Reasoning Approach, *Expert Systems*, *24*, 232-251.

[73] Watson, I. (1997). Case-Based Reasoning is a Methodology not a Technology. *Knowledge-Based Systems*, *12*, 303-308.

[74] Wilson, D. & Leake, D. (2001). Maintaining Case-Based Reasoners: Dimensions and Directions. *Computational Intelligence*, *17*, 196-213.

[75] Wylie, R., Orchard, R., Halasz, M. & Dube, F. (1997). IDS: Improving Aircraft Fleet Maintenance. In T. Senator, & B. Buchanan, (Eds.), *Proceedings of the Ninth Conference on Innovative Applications of Artificial Intelligence*, (1078-1085). Menlo Park, CA: AAAI Press.

[76] Yuan, F. C. & C. Chiu, C. (2009). A Hierarchical Design of Case-Based Reasoning in the Balanced Scorecard Application. *Expert Systems with Applications*, *36*, 333-342.

[77] Zeleznikow, J., Vossos, G. & Hunter, D. (1994). The IKBALS Project: Multi-Modal Reasoning in Legal Knowledge Based Systems. *Artificial Intelligence and Law*, *2*, 169-203.

[78] Zhou, J., Messermith, C. G. & Harrington, J. D. (2005). HIDES: A Computer-Based Herbicide Injury Diagnostic Expert System. *Weed Technology*, *19*, 486-491.

[79] Zhuang, Z. Y., Churilov, L., Burstein, F. & Sikaris, K. (2009). Combining Data Mining and Case-Based Reasoning for Intelligent Decision Support for Pathology Ordering by General Practitioners. *European Journal of Operational Research*, *195*, 662-675.

In: Case-Based Reasoning: Processes Suitability…
Editor: Antonia M. Leeland

ISBN: 978-1-61728-352-9

Chapter 2

APPLYING IMPROVED CASE INDEXING AND RETRIEVING USING EX-POST INFORMATION IN CORPORATE BANKRUPTCY PREDICTION

Hyun-jung Kim* and Woong-jang Lee
Deloitte Consulting, Yoido-dong, Youngdeungpo-gu, Seoul, Korea

ABSTRACT

Conventional statistic and artificial intelligence approaches to business prediction resolve classification and estimation tasks in accordance with supervised learning. To improve prediction accuracy of supervised learning approaches, a priori information such as discrete outcomes of classification is required. In a domain dealing with practical business problem however, the priori information resulting from the classification task is not generally pre-defined or unknown.

Business prediction research studies of statistical and artificial intelligence based methodologies in a domain of corporate failure prediction clearly defining the classification outcomes by whether the corporation is bankrupt or not have been widely studied. More recently Case-Based Reasoning (CBR), which extracts the most relevant case to predict the credit level of the targeted corporation, is being increasingly denoted as an alternative approach.

Although CBR is applied to an environment where the adequate and appropriate classification outcomes are not available, studies finding out generalized indexing and retrieving patterns from supervised learning with abundant data have been carried out for performance improvement of CBR. However, those approaches can be restricted in a practical situation as the underlying foundation of CBR originates from a problem-solving paradigm using case specific knowledge of past experience to solve a new problem. To overcome this limitation, we propose a hybrid approach combining CBR and Data Envelopment Analysis (DEA).

The results demonstrate that the hybrid approach combining CBR and DEA suggested an alternative methodology to weigh features with limited priori information.

* Corresponding author: Deloitte Consulting, Address: 7 Fl., Hewlett Packard Korea Bldg., 23-6 *Yoido-dong*, Youngdeungpo-gu, Seoul Korea, E-mail: charitas@empal.com, Tel: +82 2 6676 3703, Fax: +82 2 6674 8800

Further, through the comparative experiment between Multiple Discriminant Analysis (MDA) as supervised learning and our hybrid approach as unsupervised learning, the results indicate the hybrid approach performs better in bankruptcy prediction than MDA when priori information is unavailable. Additionally, key areas influencing CBR's prediction performance including weighting features, determining case base, selecting an optimal number of combining cases and comparing distance metrics are considered.

INTRODUCTION

Modeling bankruptcy prediction has long been regarded as an important research area in the academic and business community. Early research studies of bankruptcy prediction used statistical techniques such as Multiple Discriminant Analysis (MDA) [7, 8], logit [40] and probit [58].

Artificial intelligence approaches, in particular artificial neural networks, have been widely used in the last decade [27, 56, 51, 55]. SVM, which is a relatively new machine learning technique, was applied to bankruptcy prediction respectively by various researchers [45, 36, 25]. More recently, few studies have applied Case Based Reasoning (CBR) to predict bankruptcy or business failure [56, 12, 25]. Although numerous theoretical and experimental studies reported the usefulness of the CBR in bankruptcy prediction researches, there are several limitations in building the model. CBR is applied to an environment where the adequate and appropriate classification outcomes are not available, studies finding out generalized indexing and retrieving patterns from supervised learning with abundant data have been carried out for performance improvement of CBR. However, those approaches can be restricted in practical situations as the underlying foundation of CBR originates from a problem solving paradigm using case specific knowledge of past experience to solve a new problem.

Further, to build an accurate CBR system, various factors including weighing features, determining the number of nearest cases and selecting appropriate distance metric should be considered. This study investigates the effectiveness of a hybrid approach combining CBR and DEA in detecting the underlying data pattern for the corporate bankruptcy prediction tasks without a priori information. The preliminary results of this research show that the accuracy and generalization performance of the proposed approach which originated from unsupervised learning is better than MDA that of a representative approach in traditional supervised learning.

The remainder of this paper is organized as follows. Literature review provides a description of CBR, especially focusing on distance metric, feature selection, determining number of cases and weighting features. In the proposed model, a brief description of the hybrid approach combining CBR and DEA is provided with indication of the processes where DEA is employed into CBR's R^4 cycle. Research data and experiments describe 10 steps for experiments carried out in this research. Result and analysis summarizes and analyzes empirical results from the experiments designed. Finally, we discussed the conclusion and limitations for this research.

Literature Review

CBR as a stem of instance based learning of machine learning perspective is a reasoning approach used when generalized knowledge is lacking to solve new problems by remembering past similar experienced solutions and reusing the information and knowledge about these situations. In other words, CBR is a case explanation strategy mainly considered as a knowledge-light approach that is suited for domains in which no perfect domain knowledge exists. For this, CBR is distinguished from the other machine learning techniques since CBR has an inherent transparency that has particular advantages for explanations which are based on actual prior cases that can be presented to the user to provide compelling support for the system's conclusions [31]. The CBR problem-solving process consists of the well-known R^4 cycle of Retrieving the most relevant cases, Reusing the solutions of the retrieved cases, Revising the suggested solutions, and Retaining the new cases [1]. The outlined R^4 cycle gives an idea of the core factors determining the accuracy and efficiency of a CBR system including methods for maintaining and organizing the case base, case indexing and case retrieving techniques, the formalization of the similarity concept, and methods of case adaptation. Among the listed core factors of CBR, the efficient retrieval of how to measure the similarity of the cases and how to combine the similar cases based on an appropriate similarity metric is a challenging issue. In short, designing an effective and efficient case retrieval mechanism indeed highly influences a successful CBR system.

Although, there are various approaches to design retrieval mechanism, we use K-Nearest Neighbor (K-NN) as this technique is a class, simple and appealing method to address case classification problems. K-NN, a non-parametric algorithm that assesses similarity between a target case and a stored case based on their feature resemblance has been extensively used in the case retrieval phase of CBR. Traditional K-NN approach adopts an exhaustive search strategy to scan the overall case-base, and then select K prior cases which have the minimum dissimilarities with the new case from the case-base [Chiu and Tsai]. Despite of its wide adaption for case retrieval, K-NN tends to suffer from poor performance when 1) inappropriate or inadequate distance functions are applied, 2) there are very many features, 3) there are very few cases, 4) there is limited priori information such as discrete outcomes of classification, 5) the data is very noisy and 6) all features are assumed equally important. Thus, the improvement of retrieving process based on K-NN is considered as a task dealing with four areas of 1) embodying distance metric, 2) selecting features, 3) determining number of cases and 4) weighting features selected.

Distance Metric

Since distance metric influences the bias of K-NN classification, the choice of appropriate and adequate distance metric to improve accuracy and performance of classifiers is important. Although, a variety of distance metrics are available, two different metrics, linear distance metric and value difference metric, determined by the characteristics of attributes are introduced below.

1. Linear distance metrics

Linear distance metrics dealing with linear or continuous input attributes include the Minkowsky [10], Mahalanobis [37], Camberra, Chebychev, Quadratic, Correlation, and Chi-Square distance metrics [35]; the context-similarity measure [11] the Contrast model [51] and Hyperrectangle distance metric [44, 21].

Although there are various distance metrics as denoted before, the most commonly used is the Minkowski distance, which is defined as:

$$D(x,y) = \left[\sum_{f=1}^{n} |x_f - y_f|^p\right]^{\frac{1}{p}}$$

Where x and y are two input vectors (one being from a stored case, and the other an input vector to be classified), n is the number of input variables (attributes or features) in the application and *f* is an individual feature from 1 to *n*.

For the target case x and the source case y, the Minkowski distance of order p (p-norm distance) is defined as:

1 norm distance =Manhattan Distance

$$D(x,y) = \sum_{f=1}^{n} |x_f - y_f|$$

2 norm distance = Euclidean Distance

$$D(x,y) = \sqrt{\sum_{f=1}^{n} (x_f - y_f)^2}$$

Infinity norm distance = Max-norm Distance

$$D(x,y) = \max_{f=1}^{n} |x_f - y_f|$$

If choosing p = 1 (Manhattan Distance or City-Block Distance), big and small differences have the same influence on the distance measure. If choosing p = 2 instead, it gives more emphasis to big differences since the distance is squared. In case of the Max-norm (p = ∞) none of the differences should exceed a predefined difference.

None of the above distance metrics including Minkowski is designed to appropriately handle non-continuous input attributes. In the other words, using a linear distance metric with a nominal attribute which is a discrete whose values are not necessarily in any linear order makes little sense in distance measure.

To deal with applications having both continuous and nominal attributes, some researchers have proposed overlap distance metric uses different attributes distance metrics on different kinds of attributes. Overlap Metric (OM) for nominal attributes and normalized Euclidean distance for linear attributes [4, 2, 24] is defined as:

$$OM(x,y) = \sqrt{\sum_{a=1}^{m} d_a(x_a - y_a)^2}$$

Where

$$A = \begin{cases} 1 & \text{if } x \text{ or } y \text{ is unknown, else} \\ overlap(x,y) & \text{if } a \text{ is nominal, else} \\ \dfrac{|x-y|}{Man_x - Main_a} & \end{cases}$$

and function overlap is defined as :

$$Overlap(x,y) = \begin{cases} 0, x = y \\ 1, otherwis \end{cases}$$

Overlap Metric removes the effects of the arbitrary ordering of nominal values as it uses a distance of 1 between attribute values that are different, and a distance of 0 if the values are the same. This metric however loses much of the information that can be found from the nominal attribute values themselves [54].

2. Value difference metric

The Value Difference Metric (VDM), introduced by [50] is for nominal attributes. A simplified version of the VDM without feature weighting schemes defines the distance between two cases of x as target case and y as source case of an attribute a as:

$$vdm(x,y) = \sum_{c}^{C} \left| \frac{N_{a,x,c}}{N_{a,x}} - \frac{N_{a,y,c}}{N_{a,y}} \right|^q = \sum_{c}^{C} \left| P_{a,x,c} - P_{a,y,c} \right|^q$$

Where $N_{a,x}$ is the number of cases in the training set T that have value x for attribute a; $N_{a,x,c}$ is the number of cases in T that have value x for attribute a and output class c; C is the number of output classes in the problem domain; q is a constant, usually 1 or 2; and $P_{a,x,c}$ is the conditional probability that the output class is c given that attribute a has the value x, thus, $P(c|x_a)$. As can be seen from VDM equation, $P_{a,x,c}$ is defined as:

$$P_{a,x,c} = \frac{N_{a,x,c}}{N_{a,x}}$$

Where $N_{a.x}$ is the sum of $N_{a.x,c}$ over all classes; that is,

$$N_{a,x} = \sum_{c}^{C} N_{a,x,c}$$

And the sum of $P_{a.x.c}$ over all C classes is 1 for a fixed value of a and x [54].

Under the $VDM_a(x,y)$, the distance between two values is low when those values have more similar classifications(i.e., more similar correlations with the output classes), regardless of what order the values may be given in. The weakness of VDM is that it is inappropriate to apply VDM directly on continuous attributes. To manage continuous attributes in VDM, it is considered to transform the continuous attributes into a somewhat arbitrary number of discrete ranges, and the treated those values as nominal values. Although this approach is advantageous when generating a large enough statistical sample for each nominal value that the P values have some significance, important information available in the continuous values is lost due to the discretization. For instance, two different values in same discrete range are considered as same even if they are on opposite ends of the range. Such effects can reduce generalization accuracy [9].

Feature selection & determining number of cases

Selecting the right set of features for classification is an important problem in designing an efficient classifier. The objective of feature selection is to identify the optimal or near-optimal features in the space of feature subsets with respect to a selected performance measure. Many successful feature selection algorithms have been suggested but finding the optimum feature subset is a difficult task. Thus, many feature selection algorithms are proposed to fine suboptimum feature set in comparably smaller amount of time [26]. Stochastic algorithms including simulated annealing [47], scatter search algorithms [34] and Genetic Algorithm (GA) [13, 48] are approaches with great interests since they are often produce high accuracy and much faster. The sequential forward/backward search [3, 14, 41], filter approaches [13, 14] and branch and bound approaches [38] are the other proposed algorithms to search suboptimum solutions.

In K-NN CBR system, the parameter K means the number of cases to combine. If K value is larger than 1, it may improve the generalization properties of the retrieval and reduce sensitivity to data noise. That is, an appropriately large K parameter may enhance the accuracy of the prediction results for CBR but if K parameter is too large then the prediction accuracy may be lower because there would be many noisy cases among the selected similar cases. Therefore, determining the optimal K parameter (i.e. the number of combining cases) for K-NN is critical task to improve retrieving accuracy.

Despite of its importance to find the optimal K parameter, there have been few studies regarding K optimization problem. [32] proposed three methods which are 1) fixing the number of cases (conventional K-NN), 2) optimal spanning method, and 3) the Mathematical Programming (MP) to optimize K parameter. They provided that MP among the proposed methods was the best performed.

[5] proposed GA applied K optimization model so called, GA-K-NN (GA-optimized K – Nearest Neighbor algorithm) and showed that GA-K-NN outperforms typical CBR algorithms.

Weighting features

Most previous studies related to the K-NN simply assumed all features of cases are equally important when evaluating the dissimilarity between cases. This bias handicapped K-NN, allowing redundant, irrelevant, and other imperfect features to influence distance computations

A feature weighted K-NN matching function is mathematically represented as follows

$$D(x,y) = \sqrt{\sum_{f=1}^{n} w_f (x_f - y_f)^2}$$

where w_f is the parameterized weight value assigned to feature f, and x is the target case, y is the source case, n is the number of attributes in each case, and f is an individual feature from 1 to n. Numeric features are normalized (i.e., by subtracting their mean and dividing by their standard deviation) to ensure they have the same range and expected impact in a feature weighted K-NN matching function. The most similar cases can be retrieved when the distance (D) between x and y is minimized.

Many researchers have investigated empirical work on the feature weighting. In the early stage of feature weighing research, [29] first presented a simple approach of combining C4.5 [42, 43] and K-NN. To determine the weights (1 and 0 respectively) they use the presence and absence of attributes in the decision tress built by C4.5 with the underlying assumption of that if the attributes are not used in the decision tree, they are regarded as irrelevant

[33] considered portions of attributes in the decision tree, and showed empirically that the weight-setting strategies they presented improve predictive accuracy on artificially generated data sets and some real-world data sets.

GA as a stochastic search technique that can search large and complicated spaces is another approach to determine weighting problem in artificial intelligence algorithms including artificial neural networks, inductive learning, and linear regression [30]. [45] proposed GA-optimization for feature weights to rate corporate bond. [17] applied GA-optimized feature weighting to a real case involving customer relationship management and [18] also used the same algorithm to the due-date assignment problem in water fabrication factory.

To optimize feature weighting [23] developed an algorithm that performs search in the space of representable similarity measures using Evolutionary Algorithms (EA) searching the fittest individual, whose corresponding similarity measure yields the minimal value of an error function on the training data from the evolution selection [49].

Entropy based weighting is another approach to assign feature weights based on entropy of the distribution values of a given attribute. Under Entropy based weighting, a low weight is set if the entropy of the values distribution (or range) respect the class values distribution is high. On the other hands, a high weight is associated to a value (or range) showing a low entropy [20, 22, 53]. [39] especially proposed a local feature weighting based entropy approach which specific weights to specific values of the attribute whereas the most entropy based weighting researches use a global weighting algorithms to compute a single weight vector for all cases.

To guide effective matching and retrieval of similar companies, relevancy of the attributes is taken into consideration. [57] uses statistical evaluations for assigning the relevancy of the attributes in the NN retrieval. The results show that NN using weights derived from statistical evaluations outperforms discriminant analysis.

[19] proposed a Weighted Feature C-means clustering algorithm (WF-C-means) to group all prior cases in the case base into several clusters to obtain objective dissimilarity definition from the adjusted feature weights. Under the WF-C-Means approach, the origin feature weights are computed according to the sum of distance between clusters in terms of features then the new weights are evaluated by adding an adjustment margin calculated by applying a common decision optimization method in linear programming theory to the origin weights.

[15] introduced CBR based SVM data mining model that combines a case based reasoning and support vector machine to predict the stock price movement. They presented stepwise regression analysis to define the most relevant factors from the set of inputs to cluster the case base into a smaller case.

From the literature review, the dominant researches in CBR are based on supervised learning which requires abundant priori information to determine feature weighting.

PROPOSED MODEL

Conventional statistic and artificial intelligence approaches to business prediction resolve classification and estimation tasks based on supervised learning which requires a priori information to improve prediction accuracy. However, in the practical business problem, the priori information resulted from the classification task is not generally pre-defined or unknown.

Although CBR is applied to an environment where the adequate and appropriate classification outcomes are not available, approaches based on supervised learning with abundant priori information to improve prediction performance of CBR have been widely studied. However, those approaches can be restricted in practical situation as the underlying foundation of CBR is originated from a problem solving paradigm using case specific knowledge of past experience to solve new problem.

To overcome this limitation, we propose a hybrid approach combining CBR and Data Envelopment Analysis (DEA). DEA is an approach to measure relative effectiveness of Decision Making Units (DMUs) using ex-post information and thus a hybrid approach is expected to develop effective methods for indexing and retrieving of CBR without a priori information.

DEA is an approach to evaluate the performance of DMUs which utilize the same inputs to produce the same outputs under different conditions. From the DEA originating study, [16] described DEA as a 'mathematical programming model applied to observational data that provides a new way of obtaining empirical estimates of relations – such as the production functions and/or efficient production possibility surfaces – that are cornerstones of modern economics'. DEA as a relatively recent data oriented approach is applied to a great variety of applications in evaluating the performances of many different contexts.

The underpinning idea of DEA is to allow each DMU to use different virtual weights, the most favorable, in calculating its relative efficiency denoted as the ratio of weighted sum of outputs to weighted sum of inputs. This can be defined as followed:

$$\text{Efficiency} = \frac{\text{weighted sum of ouputs}}{\text{weighted sum of inputs}}$$

DEA makes use of fractional and corresponding linear programs to measure the relative performance of DMUs [52]. DEA is based on the economic notion of pareto optimality. A given DMU is not efficient if some other DMU can produce the same amounts of output with less of some resources and not more of any other [6].

[16] proposed that the efficiency of a target unit j_0 can be obtained by solving the following model:

$$\text{Max } h_0 = \frac{\sum_{r=1}^{t} u_r y_{rj0}}{\sum_{i=1}^{m} v_i x_{ij0}}$$

Subject to

$$\frac{\sum_{r=1}^{t} u_r y_{rj}}{\sum_{i=1}^{m} v_i x_{ij}} \leq 1, \ j = 1, \ldots, n,$$

$$u_r, v_i \geq \varepsilon, \forall r \text{ and } i,$$

Where

y_{rj} = amount of output r from unit j,
x_{ij} = amount of input i to unit j,
u_r = the weight given to output r,
v_i = the weight given to input I,
n = the number of units,
t = the number of outputs,
m = the number of inputs,
ε = a small positive number.

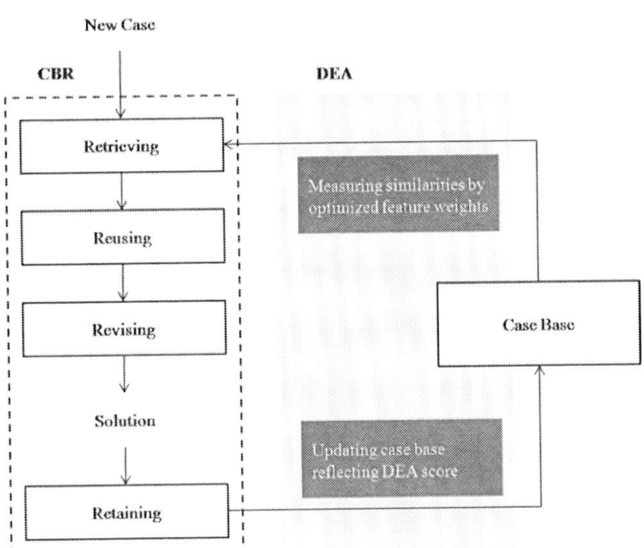

Figure 1. Hybrid Approach combining CBR and DEA

As shown above, the efficiency of a target unit j_0 can be calculated if there exist determined input and output variables for each unit without a priori information.

The following figure 1 summarizes our hybrid approach combining CBR and DEA. Within CBR's R^4 cycle, DEA is employed in Retrieving and Retaining processes specifically. In retrieving process, DEA score obtained is used to determine the optimal feature weights. When CBR case base is updated with retained cases, DEA score can be applied again to select specific retained cases.

[28] introduced a hybrid approach using DEA and CBR for housing refurbishment contractor selection and performance improvement. Among CBR's R^4, the case revision process is usually implicit for the traditional application of CBR since it may involve the user's personal judgment to adapt retrieval cases. This defect might limit the accuracy of the CBR operation, as well as decrease the robustness of applications. Thus, Juan developed a hybrid DEA-based CBR to select potential refurbishment contractors and then evaluate their performances. In the approach, DEA is integrated into the revision process of CBR, which can enhance the robustness and persuasiveness of the CBR operation.

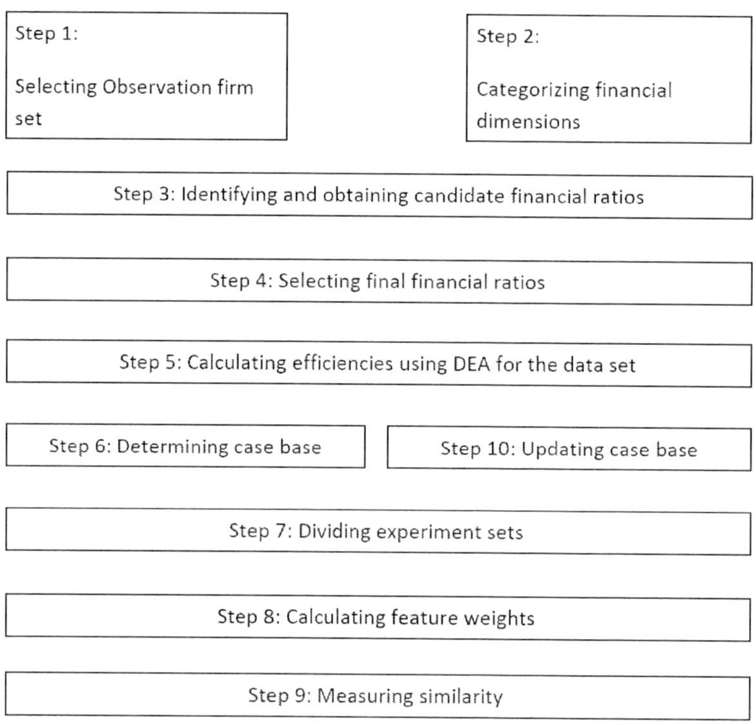

Figure 2. Flowchart of the proposed approach

RESEARCH DATA AND EXPERIMENTS

The proposed methodology consists of ten steps, as outlined in Figure 2. The first four steps are about selection of firms and financial ratios for analysis. Step 5 calculates efficiencies via DEA with the selected financial ratios for the firms selected step 1. In step 6 and 7, researcher determines case base for CBR and experimental data sets according to the DEA efficiency cutoff. Step 8 calculates feature weights through regression between DEA efficiencies and financial ratios of the case base. The similarity between cases is calculated from distance metrics of Euclidean and Manhattan with different K values in step 9. Finally, to investigate whether the prediction accuracy is enhanced when the more recent cases are added to the original case base, step 10 updates cases to the case base determined in step 6.

Step 1. Selecting Observation Firm Set

The criteria to select firms for this research are

1) External audited corporation
2) Operating in manufacturing business
3) Korea firms from the financial year of 1999 to 2005

The firms for analysis preselected upon the condition mentioned before are divided into two categories of non-default and default according to the definition of default each firm applies. The frequency of firms in non-default and default by year is presented in table 1.

Table 1. Number of companies in each year

Year	Default	Non-default	Total
1999	4	317	321
2000	18	547	565
2001	12	600	612
2002	16	709	725
2003	24	774	798
2004	8	797	805
2005	12	841	853
Total	94	4,585	4,679

Step 2. Categorizing Financial Dimensions

Using available financial information from financial statements the preselected firms published, 163 financial ratios indicating firm's performance are calculated. Then those calculated 163 financial ratios are grouped into the following 10 dimensions according to association and significance between ratios. Table 2 shows those 163 ratios by financial categories.

Step 3. Identifying and Obtaining Candidate Financial Ratios

To filter irrelevant financial ratios, firstly, univariate analysis such as T-Test, Simple Logistic, and Kolmogorov-Smirnov test (K-S Test) is applied. From the univariate analysis, statistics, coefficients and p-values for each financial ratio are obtained.

Then, to sort out the candidate financial ratios satisfying statistical significance, certain criteria denoted below is applied;

1. Ratios having same sign direction between conceptual and statistical coefficient, and
2. Ratios with p-value of univariate analysis statistics < 0.05

From the significance test above, 118 candidate financial ratios are preliminarily selected as followed in table 3.

Table 2. Number of financial ratios

Category	No. of Financial Ratios
Growth	6
Profitability	8
Debt Coverage	12
Leverage	34
Reserve Ratio	26
Activity	8
Liquidity	14
Scale	20
Cash Flow	23
Etc	12
Total	163

Table 3. Number of candidate financial ratios

Category	No. of Financial Ratios
Growth	-
Profitability	30
Debt Coverage	12
Leverage	22
Reserve Ratio	7
Activity	12
Liquidity	9
Scale	3
Cash Flow	17
Etc	6
Total	118

Step 4. Selecting Final Financial Ratios

In this step, two correlation analyses are carried out in turn to select the final candidate financial ratios. The first correlation analysis aims at reducing the data set by grouping similar ratios since the preliminarily chosen ratios are too many to be analyzed all together. Because the ratios with similar correlation coefficients in same financial category indicate similar properties, it is needed to filter the ratios to address the problem of analyzing the interrelationships among a large number of variables.

The judgmental correlation coefficient applied to filter the ratios in same financial category is 0.7 in absolute value. Finally, to determine the secondary candidate financial ratios, the following conditions are applied to the correlated ratios by categories.

1. Ratios with high accuracy rate and,
2. Ratios more explainable for analysis

The second correlation analysis is conducted over the secondary candidate financial ratios without categorization to eliminate multicollinearity between ratios and select the final financial ratios for this study. Again, the conditions mentioned before for selecting the secondary candidate ratios are applied.

The summary table for the preliminary, secondary and final financial ratios by categories is showed in table 4 below:

Table 4. Number of financial ratios selected by statistical analysis

Category	Univariate_Preliminary	Correlation 1_Secondary	Correlation 2_Final
Growth	-	-	-
Profitability	30	5	2
Debt Coverage	12	3	1
Leverage	22	6	2
Reserve Ratio	7	2	2
Activity	12	3	1
Liquidity	9	3	-
Scale	3	-	-
Cash Flow	17	6	4
Etc	6	4	2
Total	118	32	14

The definition of finally selected financial ratios is presented in table 5.

Among the final financial ratios, 5 ratios via expert opinion and researcher's decision are selected for analysis. Table 6 presents those ratios.

Table 5. Definition of financial ratios

No.	Category	Label
1	Profitability	Return on net sales
2	Profitability	EBIT(Earnings before interest and taxes) variation
3	Debt Coverage	Debt to income
4	Leverage	Debt to equity capital
5	Leverage	Net borrowing to tangible assets
6	Reserve Ratio	Earned and capital surplus to total capital
7	Reserve Ratio	Retained earnings to total assets
8	Activity	Current liabilities to sales
9	Cash Flow	Gross margin to current assets
10	Cash Flow	Ordinary profit to current liabilities
11	Cash Flow	Net cash flow to total capital
12	Cash Flow	Net cash flow to short term debt
13	Etc	Profit to value added
14	Etc	Liquidating value to sales

Table 6. The final financial ratios for analysis

No.	Category	Label
1	Profitability	EBIT variation
2	Leverage	Net borrowing to tangible assets
3	Activity	Current liabilities to sales
4	Cash Flow	Ordinary profit to current liabilities
5	Etc	Liquidating value to sales

Step 5. Calculating Efficiencies Using DEA for the Data Set

According to the DEA, the relative efficiency of each DMU is calculated by the ratio of weighted sum of outputs to weighted sum of inputs. The efficiency defined in this study is determined by the extent to which a firm is not defaulted. To evaluate a firm's soundness, the final financial ratios are separated into denominator as inputs and numerators as outputs upon their financial characteristics explain the extent of bankruptcy.

So, the first three ratios of EBIT variation, Net borrowing to tangible assets and Current liabilities to sales are classified to input variables which have positive relationship with firm's default and the rest of the final financial ratios of Ordinary profit to current liabilities and Liquidating value of sales are put into output variables which are negatively related with default. Table 7 presents the classification of the final financial ratios.

After classification of the final ratios to input and output variables, firm's DEA score of which the higher DEA score means the stronger soundness is calculated by DEA function.

Table 7. Classification for final financial ratios to calculate DEA score

No.	Category	Label	Classification
1	Profitability	EBIT variation	Input
2	Leverage	Net borrowing to tangible assets	
3	Activity	Current liabilities to sales	
4	Cash Flow	Ordinary profit to current liabilities	output
5	Etc	Liquidating value to sales	

Table 8. Case composition for Exp 1

	Case Base	Test (Predict)
Year	1999, 2000, 2001	2002, 2003

Step 6. Determining Case Base

The initial case base for this study is determined in this step. Among the data from the financial year of 1999 to 2005, firms in the year of 1999, 2000 and 2001 compose case base and firms in the rest of years from 2002 to 2003 compose the test(predict) cases. In the other word, the analysis is designed to predict firm's soundness in upcoming financial years from 2002 to 2003 by using case base composed by the data from 1999 to 2001. For research purpose, we index this experiment with case base of 1999, 2000 and 2001 as Exp 1. Table 8 summarizes the years for case base and test cases in Exp 1.

Step 7. Dividing Experiment Sets

Once, the initial case base is determined, different experiment sets based on DEA score and feature weight are considered. The following table 9 summarizes about experiment sets and table 10 presents the frequency of firms in non default and default of each experiment set.

Among the experiment sets denoted above, Exp 1_1 is control set as it does not account feature weights when similarity between cases is calculated via distance metric. The cutoffs of DEA score to constitute case base are set by researcher's judgment with considering the aim of this study.

Table 9. Summary of experiment sets

Experiment Set	Case Base Condition	Weight
Exp 1_1	Non-Defaulted firms with DEA Score > 0.5 and, Defaulted firms	No Weight
Exp 1_2	Non-Defaulted firms with DEA Score > 0.5 and, Defaulted firms with DEA Score < 0.05	Weight
Exp 1_3	Non-Defaulted firms with DEA Score > 0.5 and, Defaulted firms with DEA Score < 0.025	Weight
Exp 1_4	Non-Defaulted firms with DEA Score > 0.75 and, Defaulted firms with DEA Score < 0.025	Weight

Table 10. Number of companies in each experiment set

Set	Default	Non-default	Total
Exp 1_1	34	47	81
Exp 1_2	19	47	66
Exp 1_3	7	47	54
Exp 1_4	7	18	25
Exp 2_1	47	178	225
Exp 2_2	26	178	204
Exp 2_3	14	178	192
Exp 2_4	14	112	126

Table 11. Summary of determination for possible case combinations

When K=2				
Possible Combination	K2_unknown	K2_Non Default	K2_Default	
D*, D	Default	Default	Default	
D, N**	Unknown	Non Default	Default	
N, N	Non Default	Non Default	Non Default	
When K=3				
Possible Combination	K3_unknown	K3_Non Default	K3_Majority	K3_Default
D, D, D	Default	Default	Default	Default
D, D, N	Unknown	Non Default	Default	Default
D, N, N	Unknown	Non Default	Non Default	Default
N, N, N	Non Default	Non Default	Non Default	Non Default

* Default

** Non Default

After building experiment sets, sub-experiment sets account K-value in K-NN and distance metrics are considered for analysis. In this study, three K values of 1, 2, and 3 and two distance metrics of Euclidean and Manhattan are considered.

The possible combinations of cases with priori information about whether firms are defaulted or not and the judgments for each combination are summarized as followed in table 11.

As denoted above, sub experiment sets by K-values except K=1 (because there is no case combination with K=1) can be sorted into 4 categories of Unknown, Non Default, Default, and Majority (when K=3 only). A category 'Unknown' does not decide about vague combinations. Categories 'Non Default' and 'Default' determine possible combinations upon the tightness toward bankruptcy. Finally, category 'Majority' applies majority voting decision on default prediction.

Step 8. Calculating Feature Weights

Once, the experiment sets are determined as mentioned in step 7, the linear programming to figure out efficiency ratio via defining DEA weights is carried out. According to our hybrid approach combining DEA and CBR to find out optimal feature weights, the DEA scores obtained from linear programming are set as Y value and the final financial ratios per firms in each experiment set are placed as X value in regression analysis. As a result of regression analysis, feature weights for the final financial ratios per experiment sets are calculated.

Step 9. Measuring Similarity

This step is the final moment in our proposed methodology. The feature weights obtained through the regression analysis in Step 8, apply to the distance metrics of Euclidean and Manhattan we proposed for this study to measure similarities indicating which cases are the most similar to the target cases.

Step 10. Updating Case Base

In real world situations where CBR is applied, the issue of case maintenance becomes more critical. Unscreened case base growth influences to CBR performance being worsen as retrieval efficiency degrades and inconsistent or irrelevant cases become difficult to be detected.

For case maintenance issue in CBR, we firstly processed steps from 1 to 9 to measure similarities using the hybrid approach combining CBR and DEA. From the research, we concluded the Exp 1_3 is the best performance set. To simulate updating case base, firms in 2002 and 2003 are added to the case base of Exp 1_3 to establish the updated initial case base.

Table 12. Case composition for Exp 2

	Case Base	Test (Predict)
Year	Exp 1_3 from Exp 1 and 2002, 2003	2004, 2005

Once the updated initial case base is set, the same processes from step 1 to 9 denote above are carried out to research whether the updated case base with recent cases, in this study it will be cases in 2002 and 2003, performs better than Exp 1 to predict firms in 2004 and 2005. For research purpose, we index this experiment of updating case base with case base as Exp 2. Table 12 summarizes the years for case base and test cases in Exp 2 and the frequency of firms in non default and default of each experiment set is presented in table 10.

RESULT AND ANALYSIS

To overcome the limitation of supervised learning which needs abundant priori information to improve prediction accuracy, we suggest DEA measuring relative effectiveness of DMUs using ex-post information to combine to develop effective approach for CBR without a priori information. In our proposed hybrid approach combining DEA and CBR, DEA is applied to build up the appropriate CBR case base and more importantly to investigate the optimal feature weight using predetermined CBR case base to measure similarities between case base and target cases.

To investigate research purpose, we designed, as stated before in the proposed methodology, Exp 1 and 2 upon the initial case base is updated. The experiment results are summarized in table 13.

The results in table 13 show prediction performance indicators consisting of prediction error and prediction accuracy with respect to various K-values and distance metrics. Prediction error is divided into two types of error noted as type 1 and type 2 error. Type 1 error indicates fault prediction where default firms are predicted as non-default and type 2 error applies to the opposite case to type 1 error. Prediction accuracy intuitively means the correct prediction to a firm. Thus, the lower type 1, 2 errors and the higher accuracy conclude the better prediction performance.

Before to advance for further analysis on result, careful interpretation on sub-experiment sets of K2_Non Default and K3_Non Default where tight default determination applied are needed. As the table shows, those sub-experiments' prediction performances in terms of type 2 error and accuracy are relatively higher than the other sub-experiment sets in each experiment. This superior prediction performance however is resulted by the relatively large number of non-defaulted firms included in prediction cases since those sub-experiment sets apply tight default determination.

In contrast to the tight default determination applied sub-experiment sets, K2_Default and K3_Default with relatively lose default determination on possible predict combination show better prediction performance in terms of type 1 error which mean the other prediction performance indicators of type 2 error and accuracy rate show inferior results. The lower rate of type 1 error in those sub-experiment sets thus, should be analyzed as trade off with K2_Non Default and K3_Non Default, where the relatively large number of defaulted firms included in prediction cases relationship.

From the initial result analysis, the following analysis is based on experiments excluding the results of sub-experiments of K2_Non Default, Default and K3_Non Default and Default as those prediction results are somewhat distorted by determination on default.

Experiment 1

The Exp 1 concludes four results. The first is that experiments with setting feature weighting are better in terms of prediction performance than non feature weighting. The second result is analyzed in context of establishing case base for CBR.

Table 13. Classification accuracies and error rate (%) of various parameters on experiment sets

		Distance Metric							Distance Metric					
		Euclidean			Manhattan				Euclidean			Manhattan		
K value	Set	Type1	Type2	Accuracy	Type1	Type2	Accuracy	Set	Type1	Type2	Accuracy	Type1	Type2	Accuracy
K1	Exp1_1	5.0	67.3	34.3	5.0	68.5	33.2	Exp2_1	0.0	60.0	40.7	0.0	62.1	38.6
K2_Unknown		2.6	70.2	31.7	2.6	70.7	31.3		0.0	57.9	43.0	0.0	60.8	40.1
K2_Non Default		5.0	63.5	38.0	5.0	63.1	38.4		0.0	46.1	54.5	0.0	48.2	52.4
K2_Default		2.5	73.1	28.8	2.5	73.8	28.0		0.0	66.5	34.3	0.0	68.9	31.9
K3_Unknown		2.6	71.1	31.0	2.6	71.2	30.9		0.0	59.6	41.4	0.0	61.5	39.5
K3_Non Default		7.5	58.5	42.8	5.0	58.7	42.7		0.0	43.6	56.9	0.0	44.7	55.8
K3_Majority		2.5	69.1	32.6	2.5	70.0	31.8		0.0	51.8	48.8	0.0	55.7	44.9
K3_Default		2.5	76.2	25.7	2.5	76.3	25.7		0.0	70.5	30.4	0.0	72.0	28.9
K1	Exp1_2	7.5	49.2	51.9	7.5	47.3	53.7	Exp2_2	10.0	23.2	77.0	10.0	25.9	74.2
K2_Unknown		7.5	53.4	47.9	5.1	49.5	51.8		5.9	22.3	77.9	5.6	24.3	75.9
K2_Non Default		7.5	46.9	54.2	7.5	42.3	58.6		20.0	20.8	79.3	0.0	21.0	79.1
K2_Default		7.5	59.1	42.2	5.0	56.9	44.5		5.0	27.6	72.7	5.0	34.6	65.8
K3_Unknown		5.1	54.6	46.9	5.1	48.6	52.8		5.9	19.9	80.2	0.0	21.8	78.4
K3_Non Default		7.5	45.3	55.7	7.5	38.5	62.3		20.0	16.9	83.1	0.0	16.7	83.2
K3_Majority		7.5	57.1	44.2	5.0	53.3	47.9		15.0	24.7	75.5	10.0	29.4	70.8

Table 13. (Continued)

		Distance Metric							Distance Metric					
		Euclidean			Manhattan				Euclidean			Manhattan		
K value	Set	Type1	Type2	Accuracy	Type1	Type2	Accuracy	Set	Type1	Type2	Accuracy	Type1	Type2	Accuracy
K3_Default	Exp1_3	5.0	62.4	39.1	5.0	59.3	42.2	Exp2_3	5.0	32.1	68.2	0.0	40.1	60.4
K1		7.5	43.1	57.8	7.5	45.4	55.5		25.0	13.6	86.3	15.0	16.4	83.6
K2_Unknown		7.7	41.3	59.8	5.6	34.6	66.4		23.5	10.0	89.9	18.8	8.1	91.8
K2_Non Default		10.0	33.2	67.4	15.0	24.4	75.8		35.0	9.3	90.4	0.0	6.8	92.8
K2_Default		7.5	52.7	48.5	5.0	53.9	47.4		20.0	16.2	83.8	15.0	22.6	77.4
K3_Unknown		7.7	35.6	65.5	5.9	31.6	69.3		21.4	9.1	90.8	8.3	7.8	92.2
K3_Non Default		10.0	23.7	76.6	20.0	20.2	79.8		45.0	8.2	91.4	0.0	6.0	93.5
K3_Majority		7.5	41.7	59.2	10.0	33.3	67.3		25.0	10.5	89.3	30.0	10.2	89.6
K3_Default		7.5	57.0	44.3	5.0	56.2	45.2		15.0	18.3	81.8	5.0	28.9	71.4
K1	Exp1_4	10.0	29.9	70.7	5.0	47.6	53.5	Exp2_4	30.0	14.5	85.3	20.0	18.1	81.8
K2_Unknown		5.4	45.8	55.9	2.6	52.0	49.9		18.8	10.7	89.2	18.8	11.4	88.5
K2_Non Default		12.5	26.1	74.3	7.5	34.7	66.1		35.0	10.0	89.7	0.0	9.6	90.1
K2_Default		5.0	69.1	32.6	2.5	68.0	33.7		15.0	16.2	83.8	15.0	25.3	74.8
K3_Unknown		3.1	35.8	66.1	0.0	49.9	52.1		15.4	10.3	89.7	7.1	11.5	88.6
K3_Non Default		22.5	12.5	87.2	15.0	27.1	73.2		45.0	8.7	90.8	0.0	8.4	91.3
K3_Majority		5.0	66.7	34.9	5.0	60.1	41.3		25.0	11.3	88.5	25.0	12.7	87.2
K3_Default		2.5	77.5	24.5	0.0	72.8	29.1		10.0	23.8	76.4	5.0	35.2	65.2

Among experiment sets expl_1 to expl_4, expl_3 consisting case base of non-defaulted firms with DEA score > 0.5 and, defaulted firms with DEA score < 0.025 shows the best prediction performance with an exception expl_4 with K-value 1 and Euclidean metric which indicates the better prediction performance in type 2 error and accuracy rate.

Regarding determining the optimal K-value, it is hard to distinguish which K-value is the optimum since the performance indicators by K-values are relative to each other. Despite of the relativity between K-values, we conclude that sub-experiment sets K2_Unknown and K3_Unknown should be excluded from the result analysis nevertheless they show favorable prediction performance because they produce an issue of how to determine the vague unpredicted combinations. In short, K1 and K3_Majority would be regarded as the reasonable sub experiment sets though they are not the optimum K-values.

Lastly, the performance difference between distance metrics, Euclidean and Manhattan, is not distinguishable in this study as they result comparative performance indicators by sub experiment sets.

Experiment 2

Exp 2 is designed to investigate how the updated CBR case base influences prediction performance. In Exp 2, CBR case base is updated with firms in 2002 and 2003 to the initial CBR case base composed from expl_3.

There are two purposes for Exp 2. The First is to investigate the four general results as done in Exp 1 above. And more importantly, observing how a case base updated with cases in near future to case base influences on prediction performance, so called recency effect, is the second purpose.

In Exp 2, the four results regarding on prediction performance by setting feature weighting, establishing case base, determining optimal K-value and selecting appropriate distance metric that are all presented in Exp 1, show conclusions as similar as Exp 1.

Experiment result according to the second purpose is presented at table 14.

From the table 14, two findings are noticeable. Firstly, prediction performance is better for the case base in Exp 2 which is updated with cases in 2002 and 2003 than case base of Exp 1 which is not updated with cases in 2002 and 2003. Secondly, prediction performance is better for the cases in nearer future. In the other words, performance indicators for cases in 2004 are generally better than those for 2005.

To sum up, regarding reflecting recency effect in CBR, improved prediction performance would be achieved when case base is composed with cases which are timely close to target cases.

Table 14. Classification accuracies and error rate (%) of various case bases

			Distance Metric							Distance Metric					
			Euclidean			Manhattan				Euclidean			Manhattan		
K value	Year	Case base	Type1	Type2	Accuracy	Type1	Type2	Accuracy	Case base	Type1	Type2	Accuracy	Type1	Type2	Accuracy
K1	2004	Exp1_1	0.0	68.5	32.2	0.0	69.1	31.6	Exp2_1	0.0	59.8	40.7	0.0	62.7	37.9
	2005		0.0	69.1	31.9	0.0	69.3	31.7		0.0	60.2	40.7	0.0	61.6	39.3
K2_ Unknown	2004		0.0	70.7	30.0	0.0	70.9	29.8		0.0	57.7	43.0	0.0	60.4	40.3
	2005		0.0	71.2	29.9	0.0	71.8	29.4		0.0	58.1	43.0	0.0	61.1	39.9
K3_ Unknown	2004		0.0	70.5	30.2	0.0	70.9	29.8		0.0	51.7	48.8	0.0	55.2	45.3
	2005		0.0	71.0	30.0	0.0	71.5	29.5		0.0	52.0	48.8	0.0	56.2	44.5
K3_ Majority	2004		0.0	71.7	29.2	0.0	71.7	29.1		0.0	59.2	41.7	0.0	60.4	40.4
	2005		0.0	73.8	27.5	0.0	73.4	27.9		0.0	60.0	41.1	0.0	62.6	38.6
K1	2004	Exp1_2	0.0	50.8	49.7	0.0	47.9	52.5	Exp2_2	12.5	21.5	78.6	25.0	24.2	75.8
	2005		0.0	53.3	47.5	0.0	48.9	51.8		8.3	24.9	75.4	0.0	27.6	72.8
K2_ Unknown	2004		0.0	55.7	45.0	0.0	51.2	49.4		14.3	20.5	79.5	14.3	21.8	78.3
	2005		0.0	57.2	43.7	0.0	52.1	48.8		0.0	24.0	76.3	0.0	26.6	73.8
K3_ Unknown	2004		0.0	58.3	42.2	0.0	55.0	45.6		12.5	23.0	77.1	25.0	27.5	72.5
	2005		0.0	60.2	40.7	0.0	55.5	45.3		16.7	26.3	73.9	0.0	31.3	69.2
K3_ Majority	2004		0.0	57.8	42.9	0.0	50.8	49.7		14.3	18.1	81.9	0.0	19.4	80.8
	2005		0.0	59.0	42.0	0.0	53.5	47.4		0.0	21.7	78.6	0.0	24.1	76.3

Table 14. (Continued)

			Distance Metric							Distance Metric					
			Euclidean			Manhattan				Euclidean			Manhattan		
K value	Year	Case base	Type1	Type2	Accuracy	Type1	Type2	Accuracy	Case base	Type1	Type2	Accuracy	Type1	Type2	Accuracy
K1	2004	Exp1_3	12.5	46.7	53.7	12.5	47.2	53.2	Exp2_3	25.0	12.2	87.7	25.0	15.1	84.8
	2005		0.0	47.7	53.0	0.0	49.2	51.5		25.0	14.9	85.0	8.3	17.7	82.4
K2_ Unknown	2004		0.0	45.1	55.3	16.7	34.3	65.9		28.6	8.6	91.2	28.6	7.3	92.5
	2005		0.0	47.5	53.2	0.0	39.9	60.8		20.0	11.2	88.7	11.1	8.9	91.0
K3_ Unknown	2004		12.5	43.7	56.6	25.0	35.9	64.2		25.0	9.3	90.6	25.0	9.5	90.3
	2005		0.0	46.0	54.6	8.3	37.9	62.5		25.0	11.7	88.2	33.3	10.8	88.9
K3_ Majority	2004		0.0	36.4	64.0	16.7	31.5	68.7		33.3	7.7	92.1	20.0	7.2	92.7
	2005		0.0	41.3	59.5	0.0	37.8	62.9		12.5	10.4	89.5	0.0	8.3	91.8
K1	2004	Exp1_4	12.5	31.1	69.1	12.5	49.6	50.8	Exp2_4	25.0	13.3	86.6	25.0	17.2	82.7
	2005		0.0	34.1	66.4	0.0	50.1	50.6		33.3	15.6	84.2	16.7	19.0	81.0
K2_ Unknown	2004		0.0	48.7	52.0	0.0	54.6	46.1		28.6	9.5	90.3	28.6	10.7	89.1
	2005		0.0	53.0	48.2	0.0	58.2	42.9		11.1	11.8	88.2	11.1	12.0	88.0
K3_ Unknown	2004		0.0	68.6	32.0	0.0	62.5	38.1		25.0	10.5	89.3	25.0	11.5	88.3
	2005		0.0	70.4	30.6	0.0	64.4	36.5		25.0	12.0	87.8	25.0	13.8	86.0
K3_ Majority	2004		0.0	36.8	64.1	0.0	52.1	48.6		20.0	9.0	90.9	16.7	10.3	89.6
	2005		0.0	39.6	61.5	0.0	56.5	44.7		12.5	11.5	88.5	0.0	12.6	87.5

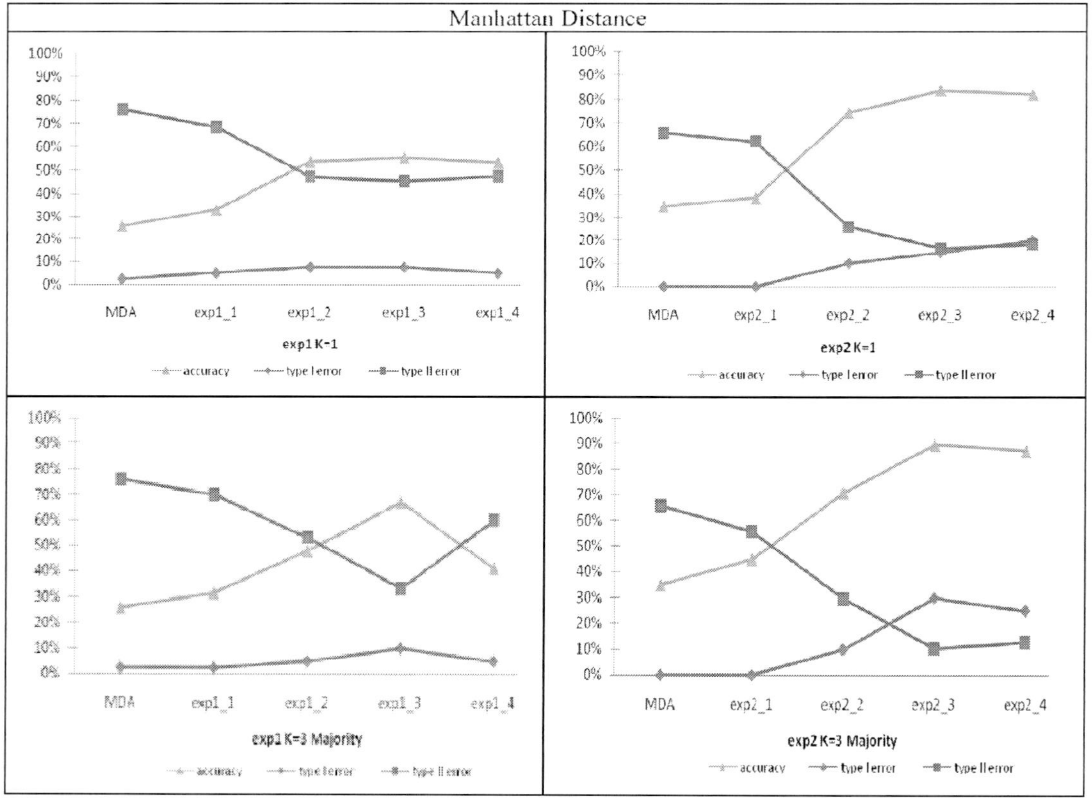

Figure 3. Continued

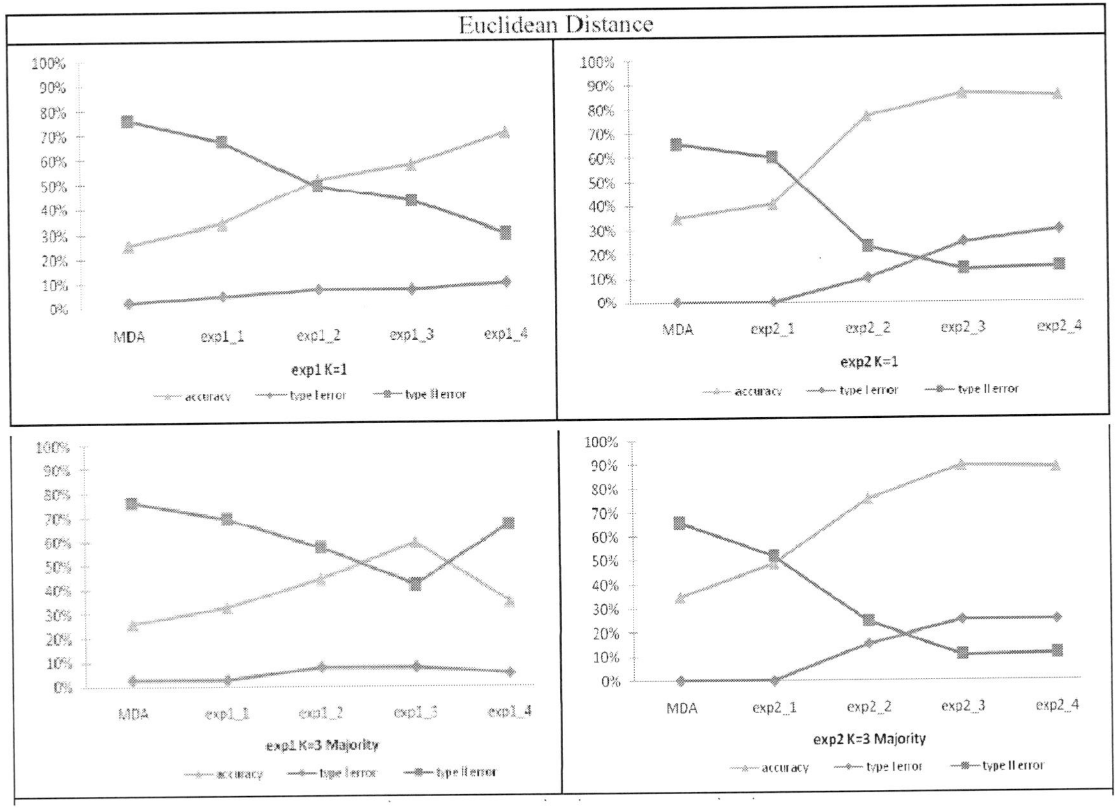

Figure 3. Supervised vs Unsupervised

Unsupervised vs. Supervised

Our hybrid approach combining CBR and DEA was initiated to deal with an environment where the abundant classification outcomes are not available so we could overcome the limitation of approaches based on supervised learning. For comparative study to provide perspectives in prediction performance, we adapted MDA which is one of the representative methodologies in supervised learning to find out default pattern recognition in the domain of bankruptcy. From the experiments, we conclude that the hybrid approach of unsupervised learning performs better than MDA of supervised learning. Summarized results are depicted in figure 3.

CONCLUSION

In the research, we aimed to present an alternative methodology to overcome the limitations the statistical supervised learning have in its application due to the violation of multivariate normality assumptions for independent variables which frequently occur in financial data. Our proposed approach employs DEA into CBR to index and retrieve similar cases with the priori information resulted by classification task is not pre-defined or unknown.

From the designed experiments, findings can be summarized in two key conclusions. Firstly, the hybrid approach combining CBR and DEA suggested an alternative methodology to weigh features with limited priori information. Secondly, through the comparative experiment between MDA as supervised learning and our hybrid approach as unsupervised learning, the results indicate the hybrid approach performs better in default prediction than MDA when limited priori information is available.

Additionally, key areas influencing CBR's prediction performance including weighting features, determining case base, selecting optimal K-value and comparing distance metrics are considered throughout Exp 1 and 2. In Exp 2, especially, the case base is updated with cases timely near to target cases to investigate the recency effect in case base maintenance problem. The results show that prediction performance for target cases which are close in time to the case base is better.

Areas where our hybrid approach would be applicable are various in real world situations. For instance, when financial institutions adapt International Financial Reporting Standards (IFRS), financial accounts constituting financial statements will be integrated, created, or eliminated to generate converged financial data. Under IFRS, financial institutions need to rebuild a bankruptcy prediction model in the credit rating system but the financial data before IFRS is difficult to be converged to meet IFRS requirements. So, due to the limited financial data under IFRS, bankruptcy prediction modeling with abundant financial data based on supervised learning is difficult. From this perspective, alternatives dealing with data limitation need to be initiated and thus our hybrid approach can be addressed as one of the potential alternatives.

There are several limitations in our research. Prediction performance in designed experiments is delivered by optimal feature weighting based on constituting an appropriate case base. In other words, it is not clear which is the determinative factor for prediction performance between appropriate case base and weighted feature. The areas where the expert's opinion or researcher's decision is applied including selecting final financial ratio and determining cutoffs of DEA scores to constitute case base are other limitations that need further research. We believe that the potential is great for further research with optimizing approaches as ways to improve the performance of the applications.

REFERENCES

[1] Aamod, A. & y Plaza, E. (1994). Case-based reasoning: Foundational Issues, Methodological Variations, and System Approaches. AI Communications, 7 (1).

[2] Aha, David W. (1992). Tolerating noisy, irrelevant and novel attributes in instance-based learning algorithms. *International Journal of Man-Machine Studies, 36*, 267-287.

[3] Aha, D. W. & Banker, R. L. (1996). A comparative evaluation of sequential feature selection algorithms. In: fisher, D., Lenx, J. -H. (eds.), *Artificial Intelligence and Statistics*, Springer-Verlag, New York.

[4] Aha, David W., Dennis Kibler & Marc K. Albert. (1991). Instance-Based learning algorithms. *Machine Learning*, 6, 37-66.

[5] Ahn, H. C. & Kim, K. J. (2008). Using genetic algorithms to optimize nearest neighbors for data mining. *Annals of Operations Research. 163,* 5-18.

[6] Al-Shammari, M. (1999). Optimization modeling for estimating and enhancing relative efficiency with application to industrial companies. *European Journal of Operational Research, 115*, 488–496.

[7] Altman, E. (1968). Financial Ratios, Discriminant Analysis and the Prediction of Corporate Bankruptcy. *The Journal of Finance, 23*, 589-609.

[8] Altman, E. (1983). *Corporate Financial Distress - A Complete Guide to Predicting, Avoiding and Dealing with Bankruptcy*. John Wiley, New York.

[9] Bao, Yongguang., Ishii, Naohiro. & Du, Xiaoyong. (2004). *Combining multiple k-nearest neighbor classifiers using different distance functions.* Springer Verlag, Berlin, 634-641.

[10] Batchelor, Bruce G. (1978). *Pattern Recognition: Ideas in Practice.* Plenum Press, New York.

[11] Biberman, Yoram. (1994). A context similarity measure. In *proceedings of the European Conference on Machine Learning* (ECML-94). Catalina, Italy, Springer Verlag, New York, 49-63.

[12] Bryant, S. M. (1997). A Case-Based Reasoning Approach to Bankruptcy Prediction Modeling. *Intelligent Systems in Accounting, Finance and Management,* 6, 195-214.

[13] Cantu-Paz, E. (2004). Feature Subset Selection, Class Separability, and Genetic Algorithms. *Genetic and Evolutionary Computation Conference*, 959-970.

[14] Cantu-Paz, E., Newsam, S. & Kamath, C. (2004). Feature selection in scientific application. In: *Proceedings of the 2004 ACM SIGKDD International Conference on Knowledge Discovery and Data Mining*, 788-793.

[15] Chang, P-C., Tsai, C-Y., Huang, C-H. & Fan, C-Y. (2009). Application of a Case Base Reasoning Based Support Vector Machine for Financial *Time Series Data Forecasting. 5755*, 294-304.

[16] Charnes, A., Cooper, W. W. & Rhodes, E. (1978), Measuring the efficiency of decision

making units, *European Journal of Operational Research 2*, 429-444.

[17] Chiu, C. (2002). A case-based customer classification approach for direct marketing. *Expert Systems with Applications, 22*, 163-168.

[18] Chiu, C., Chang, P. C. & Che, N. H. (2003). A case-based expert support system for due-date assignment in a water fabrication factory. *Journal of Intelligent Manufacturing, 14*, 287-296.

[19] Chiu, C. C & Tsai, C. (2007). *A Weighted Feature C-Means Clustering Algorithms for Case Indexing and Retrieval in Cased-Based Reasoning*. 541-551.

[20] Daelemans, W. & Van Den Bosch, A. (1992). Generalization performance of backpropagation learning on to syllabification task. In proceedings of TWLT3: Connectionism *Natural and Language Processing,* 27-37.

[21] Domingos, Pedro. (1995). Rule induction and instance-based learning: A unified approach. In *proceedings of the fourteenth international joint conference on artificial intelligence* (IJCAI-95). Morgan Karfmann, San Mateo, CA, 1226-1232.

[22] Fazil, N. (1999). Using Information Gain as Feature Weight. *8th Turkish Symposium on Artificial Intelligence and Neural Networks* (TAINN'99), Istanbul, Turkey.

[23] Gabel, T. & Riedmiller, M. (2008). Increasing Precision of Credible Case-Based Inference. *Lecture Notes in Computer Science.* 4239, 225-239.

[24] Giraud-Carrier, Christophe & Tony Martinez. (1995). An efficient metric for heterogeneous inductive learning applications in the attribute-value language. In *intelligent Systems*, vol. *1*. Edited by E. A. Yfantis. Kluwer, 341-350.

[25] Hui X. -F. & Sun J. (2006). An Application of Support Vector Machine to Companies' Financial Distress Prediction. *Lecture Notes in Computer Science, 3885*, 274-282.

[26] Jain, A. & Zongker, D. (1997). Feature selection: evaluation, application, and small sample performance. *IEEE Transactions on Pattern Analysis and Machine Intelligence 19*, 153-158.

[27] Jo, H., Han, I. & Lee, H. (1997). Bankruptcy Prediction using Case-Based Reasoning, Neural Networks, and Discriminant Analysis. *Expert Systems with Applications, 13*, 97-108.

[28] Juan, Ti-Kai. (2009). *A hybrid approach using data envelopment analysis and case-based reasoning for housing refurbishment contractors selection and performance improvement, Expert Systems with Applications 36*, 5702-5710.

[29] Kibler, D. & Aha, D. W. (1987). Learning representative examples of concepts: An initial case study. In *Proceedings of the 1987 International Workshop on Machine Learning,* Morgan Kaufmann: San Mateo, CA. 889-894.

[30] Kim, K. (2004). *Toward global optimization of case-based reasoning systems for financial forecasting. Applied Intelligence,* 21(3), 239-249.

[31] Leake, D. B. (1996). *CBR in Context: The Present and Future,* AAAI Press / MIT Press.

[32] Lee, H. Y. & Park, K. N. (1999). Methods for Determining the optimal number of cases to combine in an effective case based forecasting system. *Korean Journal of Management Research,* 27, 1239-1252.

[33] Ling, C. X., Parry, J. J. & Wang, H. (1994). *Deciding weights for IBL using C4. 5. Submitted.*

[34] Lopez, F. G., Torres, M. G., Batista, B. M., Perez, J. A. M. & Moreno-Vega, J. M. (2006) Solving feature subset selection problem by a parallel scatter search, *Eur. J.*

Oper. Res. 169, 477-497.

[35] Michalski, Ryszard S., Robert E. Stepp. & Edwin Diday. (1981). A recent advance in data analysis: Clustering objects into classes characterized by conjunctive concepts. In *progress in pattern recognition,* vol. *1*. Edited by Laveen N. Kanal and Azriel Rosenfeld. North-Holland, New York, 33-56.

[36] Min J. H. & Lee Y. -C. (2005). Bankruptcy Prediction Using Support Vector Machine with Optimal Choice of Kernel Function Parameters. *Expert Systems with Applications, 28,* 128-134.

[37] Nadler, Morton. & Smith, Eric P. (1993). *Pattern Recognition Engineering*. Wiley, New York.

[38] Narendra, P. M. & Fukunaga, K. (1977). A branch and bound algorithms for feature subset selection. *IEEE Trans. Comput. 26* (9), 917-922.

[39] Nunez, H., Sanchez, M. & Cortes., U. (2003). Improving Similarity Assessment with Entropy-Based Local Weighting. *LNAI. 2689,* 377-391.

[40] Ohlson, J. Financial Ratios and the Probabilistic Prediction of Bankruptcy. *Journal of Accounting Research, 18* (1) (1980) 109-131.

[41] Pudil, P., Novovicova, J. & Kittler, J. (1994) Floating search methods in feature selection. Pattern Recognit. *Lett. 15*, 1119-1125.

[42] Quinlan, J. R. (1993). *C4. 5: Programs for Machine Learning*. Morgan Kaufmann Publishers, Los Altos, CA.

[43] Quinlan, J. R. (1986). Induction of Decision Trees. *Machine Learning, 1*, 81-106.

[44] Salzberg, Steven. (1991). A nearest hyperrectangle learning method. *Machine Learning, 6,* 277-309.

[45] Shin, K. S. & Han, I. (1999). Case-based Reasoning Supported by Genetic Algorithms for Corporate Bond Rating. *Expert Systems with Applications, 16,* 85-95.

[46] Shin K. -S., Lee T. S. & Kim H. -J. (2005) An Application of Support Vector Machines in Bankruptcy Prediction Model. *Expert Systems with Applications, 28*, 127–135.

[47] Siedlecki, W. & Sklansky, J. (1988) On automatic feature selection, Int. J. Pattern Recogn. Artif, *Intell. 2* (2), 197-220.

[48] Siedlecki, W. & Sklansky, J. (1989). A note on genetic algorithms for large-scale feature selection. *Pattern Recognition Lett. 10*, 335-347.

[49] Stahl, A., Schmitt, S. (2002). Optimizing Retrieval in CBR by Introducing Solution Similarity. In: *Proceedings of the International Conference on Artificial Intelligence* (IC-AI 2002), Las Vegas, USA. CSREA Press.

[50] Stanfill, C. & Waltz, D. (1986). Toward memory-based reasoning. *Communications of the ACM 29*, 1213-1228.

[51] Tversky, Amos. (1977). Features of similarity. *Psychological Review, 84*(4), 327-352.

[52] Wang, W. K. (2005). A knowledge-based decision support system for measuring the performance of government real estate investment. *Expert Systems with Applications, 29*(4), 901–912.

[53] Wettschereck, D. & Dietterich, T. G. (1995) An Experimental comparison of the nearest neighbor and nearest hyperrectangle algorithms. *Machine Learning, 19,* 5-28.

[54] Wilson, D. R. & Martinez, T. R. (2000). An Integrated Instance-Based Learning Algorithm, *computational Intelligence, 16*(1), 1-28.

[55] Wilson, R. L. & Sharda, R. ((1994). Bankruptcy Prediction Using Neural Networks. *Decision Support Systems, 11,* 545-557.

[56] Varetto, F. (1998). Genetic Algorithms Applications in the Analysis of Insolvency Risk. *Journal of Banking and Finance, 22*, 1421-1439.

[57] Yip, A. Y. N. (2004). Predicting Business Failure with a Case-Based Reasoning Approach, *LNAI,* 665-671.

[58] Zmijewski, M. E. (1984). Methodological Issues Related to the Estimated of Financial Distress Prediction Models. *Journal of Accounting Research, 22* (1), 59-82.

In: Case-Based Reasoning: Processes Suitability…
Editors: Antonia M. Leeland

ISBN: 978-1-61728-352-9

Chapter 3

CASE-BASED REASONING: HISTORY, METHODOLOGY AND DEVELOPMENT TRENDS

Michael Gr. Voskoglou*

Department of Transportation and Communication Management Science, National Cheng Kung University, Tainan City, Taiwan

ABSTRACT

This paper reviews the Case-Based Reasoning (CBR) approach which, over the last few years, has grown from a rather specific and isolated research area into a field of widespread interest both from academic and commercial standpoints, and has been developed into a theory of problem-solving and learning for computers and people.

More explicitly, following an introduction with the basic concepts and a brief historical background of CBR, we focus on the steps of the CBR process, the several types of the CBR methods, the applications of CBR to a wide range of domains, and on the development trends of methods, applications and research for CBR. Finally, in our conclusion section, we underline the differences between CBR and the classical rule-induction algorithms, we refer to the existing criticism for CBR methods and, summarizing the paper, we derive our final conclusion about the CBR approach.

INTRODUCTION

Case-Based Reasoning (CBR) is a recent approach to problem-solving and learning, for computers and people, that has gotten a lot of attention over the last few yearsbecause its intelligent-systems method enables information managers to increase efficiency and reduce cost by substantially automating processes such as diagnosis, scheduling and design.

Broadly construed, CBR is the process of solving new problems based on the solutions of similar past problems. The term *problem- solving* is used here in a wide sense, coherent with

* Corresponding author: E-mail: voskoglou@teipat.gr ; mvosk@tellas.gr.

common practice within the area of knowledge-based systems in general. This means that it is not necessarily the finding of a concrete solution to an application problem; it may be any problem put forth by the user. For example, justifying or criticizing a solution proposed by the user, interpreting a problem situation, generating a set of possible solutions, or generating explanations in observable data, are also problem solving situations.

In CBR terminology, a *case* denotes a problem situation. A previously experienced situation, which has been captured and learned in a way that it can be reused in the solving of future problems, is referred to as a *past case , previous case, stored case, or retained case.* Correspondingly, a *new case, or unsolved case*, is the description of a new problem to be solved. The CBR system's expertise is embodied in a collection (library) of past cases, rather than being encoded in classical rules. Each case typically contains a description of the problem plus a solution and/or the outcomes. The knowledge and reasoning process used by an expert to solve the problem is not recorded, but is implicit in the solution.

A lawyer who advocates a particular outcome in a trial based on legal precedents, or an auto mechanic who fixes an engine by recalling another car that exhibited similar symptoms, or even a physician who considers the diagnosis and treatment of a previous patient having similar symptoms to determine the disease and treatment for the patient in front of him, are using CBR; in other words, CBR is a prominent type of analogy making.

CBR is liked by many people because they feel happier with examples than with conclusions that are separated from their context. A case-library can also be a powerful corporate resource, allowing everyone in an organization to tap in the corporate library when handling a new problem. CBR allows the case-library to be developed incrementally, while its maintenance is relatively easy and can be carried out by domain experts.

CBR is often used when experts find it difficult to articulate their thought processes when solving problems. This is because knowledge acquisition for a classical knowledge-based system would be extremely difficult in such domains, and is likely to produce incomplete or inaccurate results. When using CBR, the need for knowledge acquisition can be limited to establishing how to characterize cases. Some of the characteristics of a domain that indicate that a CBR approach might be suitable include: Records of previously solved problems exist, historical cases are viewed as an asset which ought to be preserved, remembering previous experiences is useful (experience is at least as valuable as textbook knowledge), specialists talk about the domain by giving examples.

CBR's coupling to learning occurs as a natural by-product of problem solving. When a problem is successfully solved, the experience is retained in order to solve similar problems in future. When an attempt to solve a problem fails, the reason for the failure is identified and remembered in order to avoid the same mistake in the future. This process was labeled *failure-driven learning* (Schank, 1981). Thus, CBR is a cyclic and integrated process of solving a problem, learning from this experience, solving a new problem, etc. Effective learning in CBR, sometimes referred as *case-based learning*, requires a well worked out set of methods in order to extract relevant knowledge from the experience, integrate a case into an existing knowledge structure and index the case for later matching with similar cases.

The driving force behind case-based methods has, to a large extent, come from the machine-learning community, and CBR is regarded as a subfield of machine-learning. In fact, the notion of CBR does not only denote a particular reasoning method, irrespective of how the cases are acquired, it also denotes a machine learning paradigm that enables sustained learning by updating the case base after a problem has been solved.

HISTORY OF CBR

The first trails into the CBR field have come from the study of analogical reasoning (Gentner, 1983) and –further back – from theories of concept formation, problem solving and learning within philosophy and psychology (e.g. Wittgenstein, 1953, Smith and Medin, 1981, etc). For example, Wittgenstein observed that concepts, which are part of the natural world, like bird, tree, chair, car, etc, are polymorphic and, therefore, it is not possible to come up with a classical definition, but it is better to be defined by their sets of instances, or cases.

Memory is the repository of knowledge; therefore; the question is what kind of memory accounts for observed cognitive behaviors. A leading theory has been the *semantic memory* model. Psychologists have devoted much attention to this theory (Collins & Quillian, 1969; Rumelhart et al 1972; Kintsch, 1972; etc), as have Artificial Intelligence (AI) researchers (Quillian, 1968; Woods, 1975; etc), who attempted to create computer programs that model cognitive processes. The semantic memory model typically represents static facts about the world, and, therefore, this type of knowledge does not change over time. However, it was observed that this model did not account for all the data; e. g. it does not explain how knowledge is incorporated into memory and where does the information come from.

To address these and other questions, Tulvin (1972, 1983) proposed a theory of *episodic memory* as an adjunct to semantic memory. Episodic memory receives and stores information about temporally dated episodes or events. The retrieval of information from the episodic store serves as a special type of input into episodic memory and thus changes the contents of the episodic memory store.

CBR traces its roots in AI to the work of Roger Schank and his students at Yale University – U.S.A. – in the early 1980's. Schank (1975) proposed a *conceptual memory* that combined semantic memory with Tulvin's episodic memory. *Scripts* (Schank & Abelson, 1977) were proposed as a knowledge structure for the conceptual memory. The acquisition of scripts, which are analogous to Minsky's (1975) *frames*, is the result of repeated exposure to a given situation. As a psychological theory of memory scripts suggested that people would remember an event in terms of its associated script. However, an experiment by Bower et al (1979) showed that subjects often confused events that have similar scripts: e. g. one might mix up waiting room scenes from a visit to a doctor with a visit to a dentist. These data required a revision in script theory. Schank (1979, 1980) postulated a more general structure to account for the diverse and heterogeneous nature of episodic memory, called *memory organization packet (MOP)*. MOP's can be viewed as metascripts; e. g. a professional office visit MOP can be instantiated and specified for both the doctor and the dentist, thus providing the basis for confusion between these two events.

However, more important than the MOP knowledge was the new emphasis on the basic memory processes of reminding and learning. Schank proposed a theory of learning based on reminding. According to this theory, we can classify a new episode in terms of past similar cases. Schank's model of *dynamic memory* (Schank, 1982) was the basis of the earliest CBR systems that might be called case-based reasoners: Kolodner's CYRUS (1983) and Lebowitz's IPP (1983). The basic idea of Schank's model is to organize specific cases, which share similar properties, under a more general structure called a *generalized episode (GE)*. During the storing of a new case, when a feature of it matches a feature of an existing past case, a new GE is created. Thus, the organization and structure of memory is dynamic, i. e.

changes over time. Similar parts of two case descriptions are generalized into a new GE, and the cases are indexed under this GE by their different features. Concerning CYRUS, it was basically a question-answering system with knowledge of the various travels and meetings of former US Secretary of State Cyrus Vanc,e and the case memory model developed for this system has later served as basis for several other CBR systems including MEDIATOR, PERSUADER, JULIA, etc.

An alternative approach for the representation of cases in a CBR system is the *category and exemplar model,* produced by the work of Bruce Porter and his group at the University of Texas. In this model the case memory is embedded in a network of categories, cases and index pointers. Each case is associated with a category. Finding a case in the case library that matches an input description is done by combining the features of the new problem case into a pointer to the category that shares most of these features. A new case is stored in a category by searching for a matching case and by establishing the appropriate feature indices. The above model was applied first to the PROTOS system (Porter and Bareiss, 1986, Bareiss, 1989), where emphasis is given to the combination of the general with the specific knowledge obtained through the study of cases.

Another case memory model was produced by the work of Edwina Rissland and her group at the University of Massachusetts, interested in the role of precedence reasoning in legal judgments (Rissland, 1983). This work resulted in the HYPO (Ashley, 1991) and CABARET (Skalak and Rissland, 1992) systems, where cases are grouped under a set of domain-specific dimensions.

Other early significant contributions to CBR include, the *Memory-Based Reasoning (MBR)* model of Stanfill and Waltz (1988), designed for parallel computation rather than knowledge-based matching, the study of Phyllis Koton at MIT on the use of CBR to optimize performance in an existing knowledge based system resulted in the CASEY system (Koton, 1989), etc.

In Europe research on CBR was taken up a little later, to a large extend focused towards the utilization of knowledge level modeling in CBR systems. Among the earliest results was the work of Althoff , Richter and others at the University of Kaiserslautern for complex technical diagnosis within the MOLTKE system (Althoff, 1989), which lead to the PATDEX system (Richter and Weiss, 1991), and later to several other systems and methods. In Blanes, Plaza and Lopez developed a learning apprentice system for medical diagnosis (Plaza and Lopez, 1990), while in Aberdeen Sleeman's group studied the use of cases for knowledge base refinement (REFINER system, cf. Sharma and Sleeman, 1988).

At the University of Trondheim Aamodt and colleagues at Sintef studied the learning aspect of CBR in the context of knowledge acquisition and maintenance, while for problem solving the combined use of cases and general domain knowledge was focused (Aamodt, 1989) This lead to the development of CREEK system and to continued work on knowledge-intensive CBR. On the cognitive science side, significant work was done on analogical reasoning at Trinity Colledge, Dublin (Keane, 1988) and by Strube's group at the University of Freiburg the role of episodic knowledge in cognitive models was investigated in the EVENTS project (Strube and Janetzko, 1990).

Currently, the CBR activities in the USA as well as in Europe are spreading out and the number of papers on CBR in almost any Artificial Intelligence journal is rapidly growing. Germany seems to have taken a leading position in terms of active researchers and several research groups of significant activity level have been established recently. The basic ideas and the underlined theories of CBR have spread quickly to other continents as well; from Japan, India (Venkatamaran et al, 1993) and other Asian countries, there are also activity points. In Japan the interest is mainly focused towards the parallel computation approach in CBR (Kitano, 1993).

In the 1990's , interest in CBR grew in the international community, as evidenced by the establishment of an International Conference on CBR in 1995, as well as European, German, British, Italian and other CBR workshops.

We must mention also the existence of a continuously increasing number of websites that include many references and links to electronic CBR resources, such us the US Navy Research Website, the University of Kaiserslautern Website, the AI-CBR Website of the University of Salford, including a mailing list with announcements, questions and discussions about CBR, the CBR Newsletter, that originated as a publication of the Special Interest Group on CBR in the German Society for Computer Science, the Web server of the CBR Group, part of the Department of Computer Science at the University of Massachusetts at Amherst, the Website of the Artificial Intelligence Applications Institute (AIAI), part of the School of Informatics at the University of Edinburgh, the official server of the International Conference of CBR, the AI-CBR Website of the department of Computer Science at the University of Auckland, the Machine Learning Network on line Information Service, etc . Some of the above websites are listed in detail in our references section.

The Steps of the CBR Process

CBR has been formalized for purposes of computer and human reasoning as a four step process, known as the *dynamic model of the CBR cycle*. These steps involve:

1. RETRIEVE the most similar to the new problem past case, or cases.
2. REUSE the information and knowledge in that case to solve the problem.
3. REVISE the proposed solution.
4. RETAIN the parts of this experience likely to be useful for future problem-solving.

In more detail, an initial description of a problem defines a new case. This new case is used to RETRIEVE the most similar case, or cases, from the library of previous cases. The subtasks of the retrieving procedure involve: Identifying a set of relevant problem descriptors, matching the case and returning a set of sufficiently similar cases, given a similarity threshold of some kind, and selecting the best case from the set of cases returned.

Some systems retrieve cases based largely on superficial syntactic similarities among problem descriptors, while advanced systems use semantic similarities.

The retrieved case (or cases) is combined, through REUSE, with the new case into a solved case, i.e. a proposed solution of the initial problem. The reusing procedure focuses on identifying the differences between the retrieved and the current case, as well as the part of the retrieved case which can be transferred to the new case. CBR methods are implemented by retrieval methods (to retrieve past cases), a language of preferences (to select the best case) and a form of derivational analogy (to reuse the retrieved method into the current problem).

Through the REVISE process this solution is tested for success, e.g. by being applied to the real world environment, or a simulation of it, or evaluated by a teacher, and repaired, if failed. This provides an opportunity to learn from failure.

During RETAIN, useful experience is retained for future reuse, and the case base is updated by a new learned case, or by modification of some existing cases. The retaining process involves deciding what information to retain and in what form to retain it, how to index the case for future retrieval, ant integrating the new case into the case library.

The *general knowledge* usually plays a part in the CBR cycle by supporting the CBR process. This support however may range from very weak (or none) to very strong, depending on the type of the CBR method. By general knowledge we here mean general, domain-dependent knowledge, as opposed to specific knowledge embodied by cases. For example, in the case a lawyer, mentioned in our introduction, who advocates a particular outcome in a trial based on legal precedents, the general knowledge is expressed through the existing relevant laws and the correlations between them and the case of the trial. A set of rules may have the same role in other CBR cases.

While the process-oriented view of CBR presented above enables a global, external view to what is happening, a task-oriented view could be suitable for describing the detailed mechanisms from the perspective of the CBR reasoner itself. This is coherent with the task-oriented view of knowledge level modeling, where a system is viewed as an agent which has goals, and means to achieve its goals. Tasks are set up by the goals of the system and a task is performed by applying one or more methods.

Such a *task-method decomposition* of the four main steps of the CBR process to sub-steps, where related problem-solving methods are also described, is given – in the form of a decision tree - in Aamodt & Plaza, 1994 (Figure 2). The top-level task is *problem-solving* and *learning from experience* and the *method* to accomplish the task is CBR. This splits the top-level task into the four major CBR tasks: *retrieve, reuse, revise and retain*. All the four tasks are necessary in order to perform the top-level task. The retrieve task is, in turn, partitioned into the subtasks *identify features* (collect descriptors, interpret problem, infer descriptors), *search* (to find a set of past cases), *initially match* (calculate and/or explain similarity), and *select* (the most similar case). In the same manner the reuse task is partitioned into the subtasks *copy* and *adapt* (the solution of the most similar case), the revise task is partitioned into the subtasks *evaluate solution* and *repair fault*, and the retain task is partitioned into the subtasks *integrate* (rerun problem, update general knowledge, adjust indexes), *index* (generalize and determine indexes) and *extract* (relevant descriptors, solutions, justifications and solution method). All task partitions are complete, i.e. the set of subtasks is intended to be sufficient to accomplish the task.

A method specifies the algorithm that identifies and controls the execution of subtasks, and accesses and utilizes the knowledge and information needed to do this. The methods shown in Aamodt's and Plaza's scheme, which are task decomposition and control methods, are actually high level method classes, from which one or more methods should be chosen. In this sense the method set, as shown in the scheme, is incomplete, i.e. one of the methods indicated may be sufficient to solve the task in a certain particular case, several methods may be combined, or there may be other methods that can do the job. For example, for the subtask "evaluate solution" of the task "revise" the evaluation could be done, according to the current problem, either by the teacher, or in real world, or/and in model. Another possible method, which is not shown into the scheme, is to evaluate the solution through simulation. In the same way, for the subtask "repair fault" this could be a self-repair, or a user-repair, etc.

A spherical observation of the task-oriented view of CBR described above, as Aamodt and Plaza themselves accept, makes evident that their framework and analysis approach is strongly influenced by knowledge level modeling methods in general and by the Components of Expertise methodology in particular (Steels, 1990, 1993).

Earlier flowcharts illustrating the basic steps of the CBR process were produced by Riesbeck and Bain (1987), Slade (1991), Lei et al (2001) etc.

Main Types of CBR Methods

In line with the descriptive framework for CBR presented above, core problems addressed by CBR research can be grouped into five areas: Knowledge representation, retrieval methods, reuse methods, revise methods ant retain methods. In a book published by Janet Kolodner (Kolodner, 1993), a member of Schank's research team, these problems are discussed and elaborated to substantial depth, and hints and guidelines on how to deal with them are given. An overview of the main problem issues related to these five areas is also given in Aamond & Plaza (1994) with illustrating examples drawn from the systems PROTOS, CHEF, CASEY, PATDEX, BOLERO and CREEK.

a set of coherent solutions to these problems constitutes a *CBR method*

As for Artificial Intelligence in general, there are no universal CBR methods for every domain of application. The challenge in CBR is to come up with methods that are suited for problem-solving and learning in particular subject domains and for particular application environments. Thus the CBR paradigm covers a range of different methods for organizing, retrieving, utilizing and indexing the knowledge in past cases. Actually CBR is a term used both as a generic term for the several types of these methods, as well as for one such type described below, and this has lead to some confusion. Throughout this paper we are using the term CBR in the generic sense.

The main types of CBR methods are listed below:

Case-Based Reasoning

The typical CBR methods have three characteristics that distinguish them from the other approaches listed below. First, it is assumed to have a *complexity* with respect to their internal organization, i.e. a feature vector holding some values and a corresponding class is not what we would call a typical CBR description. Second, they are able to *modify*, or adapt a retrieved solution when applied in a different problem-solving context, and third they utilize *general background knowledge*, although its richness and role within the CBR processes vary. Core methods of typical CBR systems borrow a lot from cognitive psychology theories.

Analogy-Based Reasoning

This term is sometimes used as a synonymous of the typical CBR approach (Veloso, & Carbonell, 1993), however is often used also to characterize methods, that solve new problems based on past cases of *different domains* (Kedar-Cabelli, 1988, Hall, 1989), while typical CBR methods focus on single-domain cases (a form of intra-domain analogy). The major focus of study in the analogy-based reasoning has been on the *reuse* of a past case, what is called the *mapping problem*: Finding a way to transfer, or map, the solution of an identified analogue (called *source,* or *base problem*), to the present problem (called *target problem)*. More explicitly the main steps of the analogical problem- solving process include: *Representation* of the target problem, *search-retrieval* for a related problem in memory, *mapping* of the common features of the source and of the target problem and *adaptation* of the solution procedure of the source problem for use with the target problem (Novick, 1988, Voskoglou, 2003, etc).

Exemplar-Based Reasoning

In the exemplar view a concept is defined extensionally as the set of its exemplars. In this approach solving a problem is a *classification task*, i.e. finding the right class for the unclassified exemplar. The set of classes constitutes the set of *possible solutions* and the class of the most similar past case becomes the solution to the classification problem. Modification of a solution found is therefore outside the scope of this method. Characteristic examples are the paper by Kibler and Aha (1987), and the book of Bareiss (1989).

Instance-Based Reasoning

This is a specialization of exemplar-based reasoning. To compensate for lack of guidance from general background knowledge, a relatively large number of instances is needed in order to close in on a concept definition. The representation of the instances is usually simple (e.g. feature vectors), since a major focus is to study *automated learning,* with no user in the loop. An example is the work by Aha et al (1991), and serves to distinguish their methods from more intensive exemplar-based approaches.

Memory-Based Reasoning

This approach emphasizes a collection of cases as a *large memory*, and reasoning as a process of accessing and searching in this memory. The utilization of *parallel processing* techniques is a characteristic of these methods and distinguishes this approach from the others (e.g. Stanfill & Waltz, 1988, Kolodner, 1988, Kitano, 1993, etc). The *Massive Memory Architecture* (Plaza & Arcos, 1993) is an integrated architecture for learning and problem-solving based on reuse of case experiences retained in the systems memory. A goal of this architecture is the understanding and implementing the relationship between learning and problem-solving into a reflective or introspective framework: the system is able to inspect its own past behavior in order to learn how to change its structure so as to improve its future performance.

Most CBR systems make use of general domain knowledge in addition to knowledge represented by cases. Representation and use of that domain knowledge involves *integration* of the case-based method with other methods and representation of problem-solving, for instance rule-based systems or deep models like casual reasoning. The overall architecture of the CBR system has to determine the interactions and control regime between the CBR method and the other components.

For instance, the CASEY system integrates a model-based causal reasoning program to diagnose heart diseases. When the case-based method fails to provide a correct solution, CASEY executes the model-based method to solve the problem and stores the solution as a new case for future use. Another example of integrating rules and cases is the BOLERO system (Lopez & Plaza, 1993), which has a meta-level architecture, where the base-level is composed of rules embodying knowledge to diagnose the plausible pneumonias of a patient, while the meta-level is a case-based planner that, at every moment is able to dictate which diagnoses are worthwhile to consider. In the CREEK architecture, the cases, heuristic rules, and deep models are integrated into a unified knowledge structure. The main role of the general knowledge is to provide explanatory support to the case-based processes (Aamodt, 1993); rules or deep models may also be used to solve problems on their own, if the case-based method fails. This line of work has also being developed in Europe by systems like the Massive Memory Architecture and INRECA (Manago et al, 1993). In these systems, which are closely related to the multistrategy learning systems (Michalski & Tecuci, 1992), the issues of integrating different problem-solving and learning methods are essential.

5. Tools and Applications of CBR

A CBR tool should support the four main processes of CBR: retrieval, reuse, revision and retention. A good tool should support a variety of retrieval mechanisms and allow them to be mixed when necessary. In addition, the tool should be able to handle large case libraries with the retrieval time increasing linearly (at worst) with the number of cases. CBR first appeared in commercial tools in the early 1990's and since then has been sued to create numerous applications in a wide range of domains. Organizations as diverse as IBM, VISA International, Volkswagen, British Airways and NASA have already made use of CBR in applications such as customer support, quality assurance, aircraft maintenance, process planning and decision support, and many more applications are easily imaginable. At Lokheed, Palo Alto, a fielded CBR system was developed. The problem domain is optimization of autoclave loading for heat treatment of composite materials (Hennesy & Hinkle, 1992). The autoclave is a large convection oven, where airplane parts are treated in order to get the right properties. Different material types need different heating and the task is to select the parts that can be treated together and distribute them into the oven so that their required heating profiles are taking care of. A second fielded CBR system has been developed at General Dynamics, Electric Boat Division (Brown & Lewis, 1991) handling the problem of the selection of the most appropriate mechanical equipment during the construction of ships, and to fit it to its use. Most of these problems can be handled by fairly standard procedures, but some of them, referred as "non-conformances", are harder and occur less frequently. In the period December 1990 – September 1991 20000 non-conformances were handled through the prototype CBR system that was developed and the cost reduction, compared to previous costs of manual procedures, was about 10%, which amounts to a saving of $240000 in less than one year.

In general the main domains of the CBR applications include *diagnosis, help-desk, assessment, decision support, design*, etc.

More explicitly:

CBR diagnostic systems try to retrieve past cases, whose symptom lists are similar in nature to that of the new case and suggest diagnoses based on the best matching retrieved cases.

CBR diagnostic systems are also used in the customer service area dealing with handling problems with a product or service (help-desk applications), e.g. Compaq SMART system (Nguyen et al, 1993).

In the assessment processes CBR systems are used to determine values for variables on comparing it to the known value of something similar. Assessment tasks are quite common in the finance and marketing domains.

In decision making, when faced with a complex problem, people often look for analogous problems for possible solutions. CBR systems have been developed to supporting this problem retrieval process to find relevant similar problems. CBR is particularly good at querying structured, modular and non-homogeneous documents. A number of CBR decision support tools are commercially available, including k-Commerce from eGam, Kaidara Advisor from Kaidara and SMART from Illation.

Systems to support human designers in architectural and industrial design have been developed. These systems assist the user in only one part of the design process, that of retrieving past cases, and would need to be combined with other forms of reasoning to support the full design process. An early such example is Lockheed's CLAVIER, a system for laying out composite parts to be baked in an industrial convection oven (cf. Mark, 1989).

Several commercial companies offer *shells* for building CBR systems. Just as for rule-based systems shells, they enable you to quickly develop applications, but at the expense of flexibility of representation, reasoning approach and learning methods. Four such shells are reviewed in Harmon (1992): ReMind from Cognitive Systems Inc., CBR Express/ART-IM from Inference Corporation, Esteem from Esteem Software Inc., and Induce-it (later renamed to CasePower) from Inductive Solutions Inc. On the European scene Acknosoft in Paris offers the shell KATE-CBR as part of their CaseCraft Toolbox, Isoft, also in Paris, has a shell called ReCall, TecchInno in Kaiserslauten has S3-Case, a PATDEX-derived tool that is part of their S3 environment for technical systems maintenance.

Some academic CBR tools are freely available, e.g. the PROTOS system (Porter & Bareiss, 1986), which emphasized on integrating general domain knowledge and specific case knowledge into a unified representation structure, is available from the University of Texas, and code for implementing a simple version of dynamic memory, as described in Riesbeck &

Schank (1989), is available from the Institute of Learning Sciences at Northwestern University.

A book has been published by Ian Watson (1997) in which the author explains the principles of CBR by describing its origins and constructing it with familiar information disciplines such as traditional data processing, logic programming, rule-based expert systems, and object-oriented programming. Through case studies and step-by-step examples, he goes on to show how to design and implement a reliable, robust CBR system in a real-world environment. Additional resources are provided in a survey of commercially available CBR tools, a comprehensive bibliography, and a listing of companies providing CBR software and services.

6. Development Trends of CBR Methods and Applications

The development trends of CBR methods can be grouped around five main topics.

Integration with other learning methods is the first topic that forms part of the current trend in research towards multistrategy learning systems. This research aims at achieving an integration of different learning methods into a coherent framework, where each learning method fulfills a specific and distinct role in the system, e.g. case-based learning and induction as is done in MMA and INRECA systems.

Integration with other reasoning components is the second topic that aims at using the different sources of knowledge in a more thorough, principal way, like what is done in the CASEY system with the use of causal knowledge. This trend, which is very popular in Europe, emphasizes the increasing importance of knowledge acquisition issues and techniques in the development of knowledge-intensive CBR systems.

The *massive memory parallelism* trend applies CBR to domains suitable for shallow, instance-based retrieval methods on a very large amount of data. This direction may also benefit from integration with neural network methods, as several Japanese projects currently are investigated (Kitano, 1993).

The fourth trend, *method advances by focusing on the cognitive aspects*, occurs, in particular, in the follow-up work initiated on creativity (e.g. Schank & Leake, 1989) as a new focus for CBR methods. It is not just an "application type," but a way to view CBR in general, which may have significant impacts on the CBR methods in future.

Finally, concerning the fifth topic, one must notice that, as a general problem-solving methodology intended to cover a wide range of real-world applications, CBR must face the challenge of dealing with uncertain, incomplete and vague information. In fact, successfully deployed CBR systems are commonly integrated with some method to treat uncertainty, which is already inherent in the basic CBR hypothesis demanding that similar problems have similar solutions. Correspondingly, recent years have witnessed an increased interest in formalizing parts of the CBR methodology within different frameworks of reasoning under uncertainty, and in building hybrid approaches by combining CBR with methods of uncertain and approximate reasoning. *Fuzzy sets theory* can be mentioned as a particularly interesting example. In fact, even though both CBR and fuzzy systems are intended as cognitively more plausible approaches to reasoning and problem-solving, the two corresponding fields have emphasized different aspects that complement each other in a reasonable way. Thus, fuzzy set-based concepts and methods can support the various aspects of CBR, including: Case and knowledge representation, acquisition and modeling, maintenance and management of CBR systems, case indexing and retrieval, similarity assessment and adaptation, instance-based and case-based learning, solution explanation and confidence, and representation of context. On the other hand, ideas and techniques for CBR can contribute to fuzzy set-based approximate reasoning.

In a recent paper (Voskoglou, 2009) we have constructed a fuzzy model for the description of the process of learning a subject matter in general, and we have performed a classroom experiment to illustrate its application for the case of learning mathematics in particular. More explicitly, for the construction of the fuzzy model, we considered the successive steps of the learning process (Voss, 1987) as fuzzy subsets of the set U of the linguistic labels negligible, low, intermediate, high and complete success and we have applied basic principles of the fuzzy sets theory and of uncertainty theory (e.g. see Klir & Folger, 1988). According to Voss the successive steps of the learning process include: *Representation* of the stimulus input in terms of already existing knowledge, *interpretation* of the input data, *generalization* of the new knowledge to a variety of situations, and *categorization* of the generalized knowledge into the learner's knowledge structures. In the next chapter we build an analogous model to represent the steps of the CBR process (see section 3) as fuzzy subsets in U, but further study and research is needed to check the usefulness of this representation in practice.,

Notice also that in earlier papers (Voskoglou 1996a, 1996b) we have constructed a stochastic model for the description of the analogy-based reasoning process (see section 4) and the analogical transfer of knowledge. Namely, we introduced a finite Markov chain, having as steps the corresponding stages of the analogical problem-solving process (see section 4). Applying basic principles of the finite Markov chains theory (e.g. see Kemeny & Snell, 1976) we succeeded in obtaining a measure for the skills of a group of students in solving mathematical problems via the analogical problem-solving approach. In thenext chapterthe above model is adapted to represent the CBR process in general (the Markov chain in this case should have as steps the corresponding steps of the CBR process presented in section 3), and through this we succeed in obtaining a measurement for the efficiency of a CBR system in solving similar new problems.

The trends of CBR applications clearly indicate that we will initially see a lot of help-desk applications around, and these types of systems may open up for a more general coupling of CBR to information systems. The use of cases for human browsing and decision making is also likely to lead to an increased interest in intelligent computer-aided learning, training and teaching, since CBR systems are able to continually learn from and evolve through the capturing and retaining of past experiences. On the other hand, the diagnostic systems (mainly for medical purposes) and the legacy databases will continue to be some of the most common applications of CBR. AIAI, for example, at the School of Informatics of the

University of Edinburgh, has applied CBR to otherwise intractable problems such as fraud screening.

CONCLUSION

Longstanding research in Artificial Intelligence and related fields has produced a number of paradigms for building intelligent and knowledge-based systems, such as rule-based reasoning, constraint processing, or probabilistic graphical models. Being one of these paradigms, CBR has received a great deal of attention in recent years and has been used successfully in diverse application areas. The key idea of CBR is to tackle new problems by referring to similar problems that have already been solved in the past. To this end, CBR proceeds from individual experiences in the form of cases. The generalization beyond these experiences is largely founded on principles of analogical reasoning in which the cognitive concept of similarity plays an essential role. CBR emphasizes problem-solving and learning as two sides of the same coin: Problem-solving uses the results of past learning episodes, while it provides the backbone of the experience from which learning advances.

The current state of art in Europe regarding CBR is characterized by a strong influence of the USA's ideas and CBR systems, although Europe is catching up and provides a somewhat different approach to CBR, particularly in its many activities related to integration of CBR and other approaches and by its movement toward the development of application-oriented CBR systems. The basic ideas of CBR have spread quickly to other continents; from Japan, India and other Asian countries there are also activity points.

The key difference between CBR and the classical rule-induction algorithms, which are procedures for learning rules for a given concept by generalizing from examples of that concept, lies in when the generalization is made. In fact, while CBR starts with a set of cases of training examples and forms generalizations of these examples by identifying commonalities between a retrieved case and the target problem, a rule-induction algorithm draws generalizations before the target problem is even known; i.e. it performs eager generalization. In mathematics, for example, the process of proving the truth of a proposition that depends on a non negative integer by applying induction can be consolidated by generalizing a series of suitable examples, i.e. by a rule induction algorithm. On the contrary, when a concrete problem is given, the solver has simply to retrieve in memory an analogous problem solved in the past by induction and apply the same method for the solution of the given problem (CBR).

All inductive reasoning, where data is too scarce for statistical relevance, is inherently based on anecdotal evidence. Critics of CBR argue that it is an approach that accepts anecdotal evidence as its main operating principle, but without statistically relevant data for backing an implicit generalization, there is no guarantee that the generalization is correct. Our personal opinion is that the above criticism has only a theoretical base, because in practice the CBR methods give satisfactory results in most cases.

Conclusively CBR has blown a fresh wind and a well-justified degree of optimism into Artificial Intelligence in general, and knowledge based decision support systems in particular. The growing amount of on going CBR research has the potential for leading into significant breakthroughs of Artificial Intelligence methods and applications.

REFERENCES

Aamodt, A. (1989). Towards robust expert systems than learn from experience – an architectural framework, In J., Boose, B., Gaines, J. G. Ganascia, (Eds.), EKAW-89, *Third European Knowledge Acquisition for Knowledge-Based Systems Workshop*, 311-326, Paris.

Aamodt, A. (1993). Explanation-driven retrieval, reuse and learning of cases, In EWCBR-93: *First European Workshopon Case-Based Reasoning*, University of Kaiserslautern SEKI Report SR-93-12 (SFB 314), 279-284.

Aamodt, A. & Plaza, E. (1994), Case-Based Reasoning:: Foundational Issues, Methodological Variations, and System Approaches, A. I. *Communications*, *7,* no. 1, 39-52.

Aha, D., Kibler, D. & Albert, M. K. (1991). Instance-Based Learning Algorithms, *Machine Learning*, *Vol. 6 (1)*.

AI-CBR Website, htpp://www.ai-cbr.org, *Department of Computer Science*, University of Auckland.

Althoff , K. D. (1989). Knowledge acquisition in the domain of CNC machine centers:the MOLTKE approach, In J., Boose, B. Gaines, & J. G. Ganaskia, (Eds), EKAW-89, *Third European Knowledge Acquisition for Knowledge-Based Systems Workshop*, 180-195, Paris.

Artificial Intelligence Applications Institute (AIAI) Website, htpp://www.aiai.ed.ac.uk, *School of Informatics*, University of Edinburgh.

Ashley, K. (1991), Modeling legal arguments: *Reasoning with cases and hypotheticals*, MIT Press, Bradford Books, Cambridge.

Bareiss, R. (1989). Exemplar-based knowledge acquisition: A unified approach to concept representation, *classification, and learning*, Boston, Academic Press.

Bower, G., Black, J. & Turner, T. (1979). Scripts in Memory for Text, *Cognitive Psychology*, *11*, 177-220.

Brown, B. & Lewis, L. (1991). A case-based reasoning solution to the problem of redundant resolutions of non-conformances in large scale manufacturing, In R. Smith, & C. Scott, (Eds): *Innovative Applications for Artificial Intelligence*, 3, MIT Press.

CBR Group Web server, htpp://www.cs.umass.edu/~cbr/index.html

Collins, A. & Quilliam, M. (1969). Retrieval Time from Semantic Memory, *Journal of Verbal Learning and Verbal Behavior*, *8*, 240-247.

Gentner D. (1983). Structure mapping – a theoretical framework for analogy, *Cognitive Science*, *7*, 155-170.

Hall, R. P. (1989). Computational approaches to analogical reasoning: A comparative analysis, *Artificial Intelligence*, *39, no. 1*, 39-120.

Harmon, P. (1992). Case-based reasoning III, *Intelligent Software strategies*, *VIII(1)*.

Hennssy, D. & Hinkle, D. (1992), Applying case-based reasoning to autoclave loading, *IEEE Expert*, *7(5)*, 21-26.

Keane, M. (1988). *Where's the Beef? The Absence of Pragmatic Factors in Pragmatic Theories of Analogy,* In Proc., ECAI-88, 327-333.

Kedar-Cabelli, S. (1988). Analogy – from a unified perspective, In D. H. Helman, (ed), *Analogical reasoning*, 65-103, Kluwer Academic.

Kemeny, J. & Snell, J. l. (1976). *Finite Markov Chains*, Springer-Verlag, New York.

Kintsch, W. (1972). Notes on the Structure of Semantic Memory. *In Organization of Memory*, E. Tulvin, & W. Donaldson, (Eds), 247-308, New York, Academic.

Kibler, D. & Aha, D. (1987). Learning representative exemplars of concepts: An initial study, *Proceedings of the 4th International Workshop on Machine Learning*, UC- irving, 24-29.

Kitano, H. (1993). *Challenges for massive parallelism, Proceedings of the 13th Intern.* Conference on A. I., 813-834, Morgan Kaufman, Chambery, France.

Klir, G. J. & Folger, T. A. (1988). Fuzzy Sets, *Uncertainty and Information*, Prentice Hall Int., London.

Kolodner, J. (1983). Reconstructive Memory: A Computer Model, Cognitive *Science*, *7*, 281-328.

Kolodner, J. (1988). Retrieving events from case memory: A parallel implementation, Proceedings from the Case-based Reasoning Workshop, 233-249, *Morgan Kaufmann Publ.*, Clearwater Beach, Florida.

Kolodner, J. (1993). Case-Based Reasoning, Morgan Kaufmann.

Koton, Ph. (1989). Using experience in learning and problem solving, MIT, *Laboratory of Computer Science*, (Ph.D. diss.), MIT/LCS/TR-441

Lebowitz, M. (1983). Memory-Based Parsing, *Artificial Intelligence*, *21*, 363-404.

Lei, Y., Peng, Y. & Ruan, X. (2001). Applying casebased reasoning to cold forcing process planning, *Journal of Materials Processing Technology*, *112*, 12-16.

Lopez, B. & Plaza E. (1993). Case-based planning for medical diagnosis, In Z. Kmorowski, & W. Ras, (Eds), Methodologies for Intelligence Systems: 7th International Symposium (ISMIS 93), 96-105, *Lecture Notes in Artificial Intelligence 689*, Springer Verlag.

Machine learning network on line information service, htpp://www.kdubig.org/kdubig/control/index.

Manago, M., et al (1993). Induction and reasoning from cases, In ECML- European Conference on Machine Learning, *Workshop on Intelligent Learning Architectures*, Vienna.

Mark, B. (1989). *Case-Based Reasoning for Autoclave Management*, Proceedings of the Case-Based Reasoning Workshop.

Michalski, R. & Tecuci, G. (1992). *Proc. Multistrategy Learning Workshop*, George Mason University.

Minsky, M. (1975). A Framework for Representing Knowledge. *In The Psychology of Computer Vision*, P. Wilson, (Ed), 211-277, New York, McGraw-Hill.

Novick, L. R. (1988). Analogical transfer, problem similarity and expertise, Journal of Educational Psychology: *Learning, Memory and Cognition*, *14*, 510-520.

Nguyen, T., Czerwinski, M. & Lee, D. (1993). *COMPAQ QuickSource: Providing the Consumer with the Power of Artificial Intelligence*, Proceedings of the 5th Annual Conference on Innovative Applications of Artificial Intelligence, 142-151, AAAI Press, Washington DC.

Official server of the International Conference on CBR, htpp://www.iccbr.org.

Plaza, E. & Lopez de Mantaras, R. (1990), A case-based apprentice that learns from fuzzy examples, In Z., Ras, M. Zemankova, & M. L. Emrich, (Eds), *Methodologies for Intelligent System*, *5*, 420-427, North Holland.

Plaza, E. & Arcos, J. L. (1993). Reflection and Analogy in Memory-based Learning, Proc. *Multistrategy Learning Workshop*, 42-92.

Porter, B. & Bareiss, B. (1986). PROTOS: An experiment in knowledge acquisition for heuristic classification tasks, Proceedings of the 1st Intern. *Meeting on Advances in Learning (IMAL)*, 159-174, Les Arcs, France.
Quilliam, M. (1968). Semantic Memory. In *Semantic Information Processing*, M. Minsky, (Ed), 227-353, Cambridge, Mass., MIT Press.
Richter, A. M. & Weiss, S. (1991). Similarity, uncertainty and case-based reasoning in PATDEX, R. S. In Boyer, (Ed.), *Automated reasoning, essays in honour of Woody Bledsoe*, 249-265, Kluwer.
Riesbeck, C. & Bain, W. (1987). *A Methodology for Implementing CaseBased Reasoning Systems*, Lockheed.
Riesbeck, C. & Schank, R. (1989). Inside case-based reasoning, *Lawrence Erlbaum*, 1989.
Rissland, E. (1983). Examples in legal reasoning: Legal hypotheticals, *In Proceedings of the Eight International Joint Conference on Artificial Intelligence*, IJCAI, Karlsruhe.
Rumelhart, D., Lindsay, P. & Norman, D. (1972). A Process Modelfor Long-Term Memory. *In Organization of Memory*, Tulving, E. and Donadlson (Eds), 197-246, New York, Academic.
Schank, R. (1975). The Structure of Episodes in Memory. *In Representation and Understanding*, D., G. Bobrow, & A. Collins, (Eds). 237-272, New York, Academic.
Schank, R. & Abelson, R. (1977). *Scripts, Plans, Goals and Understanding*, Hillsdale, N. J., Lawrence Erlbaum.
Schank, R. (1979). *Reminding and Memory Organization: An Introduction to MOPS*, Technical Report 170, Dept. of Computer Science, Yale University.
Schank, R. (1980). *Language and Memory, Cognitive Science*, *4(3)*, Schank, R.243-284.
Schank, R. (1981). *FailureDriven* Memory, Cognition and Brain Theory, *4(1)*, 41-60.
Schank, R. (1982). Dynamic memory; *a theory of reminding and learning in computers and people*, Cambridge Univ. Press.
Schank, R. & Leake D. (1989), Creativity and learning in case-based explainer, *Artificial Intelligence*, *40*, no. 1-3, 353-385.
Sharma, S. & Sleeman, D. (1988). REFINER; a case based differential diagnosis aide for knowledge acquisition and knowledge refinement, In EWSL 88; *Proceedings of the Third European Working Session on Learning*, 201-210, Pitman.
Skalak, C. B. & Rissland, E. (1992). Arguments and cases: An inevitable twining, Artificial Intelligence and Law, *An International Journal*, *1(1)*, 3-48.
Slade, S. (1991). Case- Based Reasoning: A Research Paradigm, *Artificial Intelligence Magazine*, *12(1)*, 42-55.
Smith, E. & Medin, D. (1981). *Categories and concepts*, Harvard University Press.
Stanfill, C. & Waltz, D. (1988). The memory-based reasoning paradigm, In Case-based reasoning, Proceedings from a workshop, 414-424, Morgan Kaufmann Publ., Clearwater Beach, Florida.
Steels, L. (1990). Components of expertise, *Al Magazine*, *11(2)*, 29-49.
Steels, L. (1993). The componential framework aqnd its role in reusability, In J. M., David, J. P. Krivine, & R. Simmons, (Eds), *Secong generation expert systems*, 273-298, Springer.
Strube, G. & Janetzko, D. (1990). Epishodishes Wissen und Fallbasierte Schliessen: Aufgade fur die Wissendsdiagnostik und Wissenspsychologie, *Schweizerische Zeitschrift fur Psychologie*, *49*, 211-221.

Tulving, E. (1972). Episodic and semantic memory, In Tulving E. and Donaldson W. : *Organization of memory*, 381-403, Academic Press.

Tulving, E. (1983). *Elements of Episodic Memory*, Oxford, Oxford University Press.

Veloso, M. M. & Carbonell, J. (1993). Derivational analogy in PRODIGY, *Machine Learning, 10(3)*, 249-278.

Venkatamaran, S. et al, (1993). A rule-rule case based system for image analysis, In First European Workshop on Case-based Reasoning, *Posters and Presentations*, Vol. *II*, 410-415, University of Kaiserslautern.

Voskoglou, M. (1996a). Use of absorbing Markov chains as a measurement model for the process of analogical transfer, Int. *J. Math. Educ. Sci., Technol., 27*, 197-205.

Voskoglou, M. (1996b). An application of ergodic Markov chains to analogical problem solving, *The Mathematics Education (India)*, Vol. *XXX (2)*, 96-108.

Voskoglou, M. (2003). Analogical problem solving and transfer, Proceedings 3[d] Mediterranean Conf. Math. *Educ.*, 295-303, Athens.

Voskoglou, M. (2009). Fuzziness, or probability in the process of learning? A general question illustrated by examples from teaching mathematics, *Journal of Fuzzy Mathematics, 17(3)*, 679-686, International Fuzzy Mathematics Institute, Los Angeles.

Voss, J. F. (1987). Learning and transfer in subject matter learning: A problem-solving model, *International Journal of Educational Research, 11*, 607-622.

Watson , I. (1997). Applying Case Based Reasoning: *techniques for Enterprise Systems*, Elsevier, Burlington.

Wikipedia, the free encyclopedia: Case-based reasoning, htpp://en.wikipedia.org/wiki/Case-beased_reasoning

Wittgenstein, L. (1955). *Philosophical investigations*, 31-34, Blackwell.

Woods, W. (1975). What's a Link: Foundations for Semantic Networks. In *Representation and Undertanding,* D. Bobrow, & A. Collins, (Eds), 35-82, New York, Academic.

In: Case-Based Reasoning: Processes Suitability...
Editors: Antonia M. Leeland
ISBN: 978-1-61728-352-9

Chapter 4

A Temporal Case-Based Procedure for Cancellation Forecasting: A Case Study

Tsung-Hsien Tsai*
Department of Transportation and Communication Management Science, National Cheng Kung University, Tainan City, Taiwan

Abstract

Given that customers can reserve many days before the service day, they, not uncommon, are also allowed to cancel their reservations before using the service. Expecting the volume of cancellations helps operators effectively manipulate their resources, which in many cases are limited and also perishable such as hotel rooms and railway seats. This paper proposes a case-based predicting model for the purpose of cancellation forecasting. Under the stages of retrieving, reusing and revising, the first major contribution of this paper is on the inclusion of temporal features of curves in the stage of pattern retrieval. Temporal features such as day-of-week, recency effect and reliability of information over booking days are investigated to generate a reasonable method for case retrieval. Another contribution is the integration of a direct search algorithm for parameter estimation in the stage of revising. Hooke-Jeeves algorithm is applied to search five key parameters in the proposed predicting model. The empirical study, which uses real railway data, shows that the proposed case-based predicting model can have at least 20% improvement of MSE over pick-up and regression models, which are two popular benchmarks in practice. Similar concepts can be extended to other industries with cancellation behavior such as airlines, restaurants, hotels, rental cars, golf courses etc.

Keywords: Forecasting, Cancellation, Case-based predicting, Data mining, Revenue management, Railway demand

---

* Corresponding author: *thtsai@mail.ncku.edu.tw*, +886-6-2757575 ext. 53270 ext. 5041.

1. Introduction

In many industries, it is a common phenomenon for customers to book their desired products or services in advance, especially when the capacity is limited. For example, customers may reserve limited debuted records or railway seats many days before they actually get the record or use the service. In most situations, customers, not uncommon, may also cancel their reservations before they use the product or service with or without paying penalties. In the airline industry, cancellation fees are usually applied when passengers choose to cancel their reservations. Industries such as hotels and rental car companies allow their clients to cancel bookings without charging a penny if cancellations are done within a time period such as 24 hours before using the service. An extreme example is in the restaurant industry which restaurateurs even charge nothing for no-shows in most situations. One severe problem of cancellation is that customers who book late are denied if the number of bookings has reached the capacity. These late and denied booking clients, however, attempt to pay more for their desired services than early booking customers. Most of denials will switch to other alternative services as a consequence. A powerful solution to tackle the abovementioned problem is overbooking which operators sell more seats (higher than the capacity) to compensate the anticipated cancellations. Smith et al. (1992) indicated that without overbooking controls, American Airlines may have 15 percent spoiled seats on sold-out flights.

Revenue management (RM) is widely utilized to help operators arrange perishable capacity so that maximized revenues can be achieved. Kimes (2005) has indicated that using RM may bring 3-5% increase of revenues in the hotel, rental car, and airline industries. In RM, cancellation forecasting offers essential information to determine the volume of overbooking. The capacity and the volume of overbooking form the so-called pseudo capacity which is an important constraint while deciding the allotment of resources. As a result, the accuracy of cancellation forecasting is critically related to the performance of revenue management systems. Lee (1990) has shown the value of accurate forecasting in RM which 10% increase of predictive accuracy can bring 0.5-3% improvement of revenues for high demand flights. Ridel and Gabrys (2007) also mentioned that 10% reduction of forecast errors can bring 2-4% of additional expected revenues for airlines.

Starting from historical booking models, which bases on conventional time series perspective and utilizes solely cancellation numbers on the service day such as exponential smoothing and moving average (Weatherford and Kimes, 2003), more sophisticated models are also available in the literature. For example, ARIMA, neural networks, and machine learning algorithms are all common and potential alternatives. In M3-competition, Makridakis and Hibon (2000) have compared the performance of 24 time series techniques on 3003 time series data. These 24 tested methods can all be potential tools for cancellation forecasting.

Another important stream to predict cancellations (or no-show/stand-by/refund/exchange) is based on the use of passenger name records (PNR), which is the most detailed information collected in airline reservation systems. Four major types of features are usually extracted from PNR data for model construction: flight attributes, airport attributes, seasonal influences, and passenger attributes (Neuling et al., 2004). Flight attributes include information related to flight itself such as whether a trip is long- or short-haul and also the frequency. For example, long-haul passengers are less likely to cancel their reservations because they have planned their trips more thoroughly in advance. Another fact is that passengers are prone to make cancellation if the frequency is high since the cost of switch is going to be insignificant. Airport attributes contain variables such as the scale of origins and destinations. Cities with large population seem to have more cancellations because they have more bookings in the beginning. In addition, if passengers arrive at an airport in the suburban area, they might want to switch to other flight alternatives landing in the downtown (for the same destination). Seasonal influences are straightforward. For instance, summer is a peak season for traveling, and passengers are more likely to keep their promises and show up at the airport. If special events are held such as computer exhibitions and carnivals during the year, customers are more likely to show up as well. Passenger attributes like whether customers are frequent fliers and the purposes of their trips are also related to their attempt of cancellation. Here we only introduce several possible causes for cancellation; readers who are interested in PNR models can refer to Garrow and Koppelman (2004a) and Garrow and Koppelman (2004b) for more thorough discussions.

In PNR models, another important feature is the selection of mapping functions. Various prototypes have been proposed in the literature. Candidate models are like decision tree (Lawrence et al., 2003; Neuling et al., 2004), probability model (Lawrence et al., 2003), choice model (Garrow and Koppelman, 2004a; Garrow and Koppelman, 2004b), support vector machine (Romero Morales and Wang, 2008), neural networks (Wu and Lin, 1999), and combined models (Hueglin and Vannotti, 2001; Neuling et al., 2004; Gorin et al., 2006; Lemke et al., 2009). The merit of this stream is theoretically sound and capable of explaining customers' behavior. One weakness of PNR models is the use of abundant data which might result in high storage and construction cost. However, this problem is, in fact, mitigated because of the development of IT in the last few decades. Another disadvantage of PNR models is that some of the abovementioned passenger attributes might not be available in some industries. For instance, restaurants or railways probably only record very basic information (names and service dates) related to customers' attributes when they make reservations. In this situation, modelers run the risk of omitting important personal variables if PNR models are applied.

Advanced booking models, which utilize the built-up of cancellations over the booking period, are another stream of potential alternatives. Regression and pick up are two representative models (Gorin et al., 2006; Wickham, 1995). Both models map the relationship between the final number of cancellations on the day of service and the number of the accumulated cancellations during the booking period. For example, regression can find the linear relationship between cancellations on the service day and cancellations 7 days ago. This linear relationship can then be applied for forecasting purposes. An advantage of these two linear models is simple to use and easy to understand; however, they might not be able to obtain satisfactory performance when situations become complicated.

Sophisticated uses of the concept based on advanced booking models are also available in the literature. Schwartz and Hiemstra (1997) introduced the concept of booking curves and proposed a model based on the similarity of curves for the prediction of hotel bookings. In a preliminary study, Tsai (2009) has designed a case-based predicting framework which considers not only the patterns of booking curves but also temporal influences for railway arrival forecasting. Temporal influences in his model are the reliability of information over the booking period and also the recency effect. In addition, an adapting term was also incorporated to dynamically capture the status of current bookings in his model.

The aim of this study is to extend the study based on Tsai (2009) for cancellation forecasting. We will propose a case-based predicting model which is a refined version of Tsai's model by further incorporating the influence of day-of-week periodicity. More distinctively, the proposed model is equipped with the mechanism of parameter estimation. This innovation will make the model applicable in different data situations. Since this proposal is based on the concept of advanced booking models, the built-up of cancellations is the only information needed. Comparing with PNR models, no other variables or attributes are required in the proposed model which is suitable for the situation where influential variables are difficult to collect or hard to measure.

We will first brief the idea of cancellation curves and extract data features from the collected railway data. After that, we will present the proposed model and also other conventional methods, i.e., regression and pick up models. A case study is then implemented to show the potential of the proposal over the benchmarks. Conclusions and future extensions are also rendered.

2. Cancellation Curves

In this section, we draw cancellation curves of a studied railway service, and important features are extracted. The studied data series is an origin-destination pair in the west coast of Taiwan. The reason for the choice is because of its prosperous demand. Since demand is high, we expect to see strong cancellations in this studied service. We currently concentrate on one single series in this study to demonstrate how to design the proposed model based on extracted features.

For a departure day, passengers can start booking two weeks in advance. After booking, passengers can make their final decision of either paying for tickets or cancelling reservations within 5 days (including the booking day). If they fail to make one of the above two decisions, the reservation system will cancel the reservation automatically and mark the record as a failure. If passengers are identified to have five failures, their rights of reserving tickets will be suspended for the next three months. If a booking date is less than five days before the departure, passengers have to make their decisions at least one day before the departure. Based on these rules, we can generate cancellation curves which record the number of accumulated cancellations at each data collection point (DCP) for the studied service. The data was collected from a railway company in Taiwan from December, 2004 to December, 2005. A DCP is defined as a booking day dated before the departure, and 15 DCPs are then used correspondingly in this study. DCP(0) records the final number of cancellations. Figure 1 illustrates several cancellation curves of the studied train service; each curve represents a specific departure day. Initially, we can see that some curves have high reservations in the beginning of the booking period; some have strong demand at the end of the booking period; and also some have stable growth of bookings over the whole period. We will extract data features based on the established cancellation curves in the next section.

2.1. Canceling Patterns before Departure

Based on the distribution of cancellation curves in Figure 1, it is straightforward to see that all cancellation curves are climbing up monotonically from DCP(14) to DCP(1) since cancellations are accumulated in nature over DCPs. In addition, the numbers at DCP(0) and DCP(1) should be equal because cancellation is only allowed at least one day before departure. Here an important feature derived from the fact of accumulation is the reliability of information at each DCP. In the beginning of the booking period, the number of cancellations is usually small and unreliable to predict cancellations at DCP(0). As DCPs are approaching the departure day, information becomes more and more reliable and ultimately reaches the final number of cancellations. The abovementioned observation advocates the following two phenomena. First, two cancellation curves which are different in the beginning of the booking period should be considered similar if their cancellation numbers become close as DCPs are approaching the departure day. Figure 2 portrays the described scenario by using a real example. On the contrary, two cancellation curves which are very close in the beginning of the booking period may turn out to be totally different as DCPs are approaching the departure day, as shown in Figure 3.

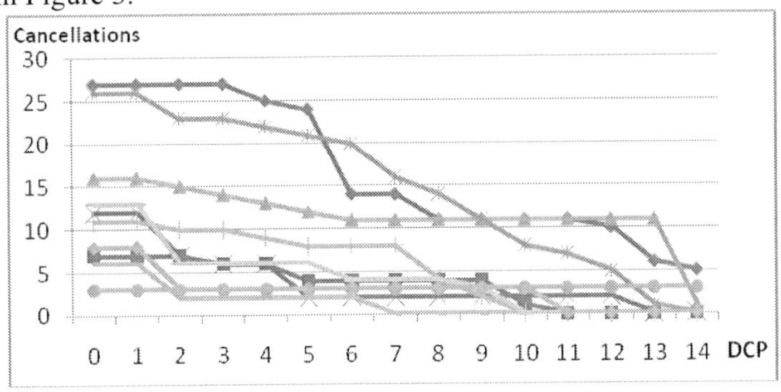

Figure 1. Examples of cancellation curves

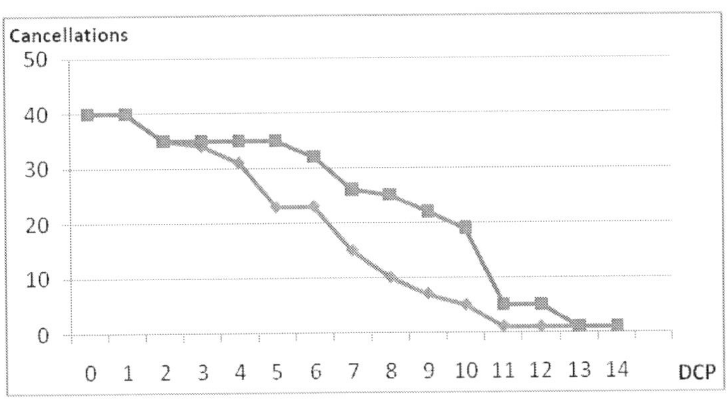

Figure 2. Two curves with different patterns in the beginning but similar cancellation numbers at DCP(0)

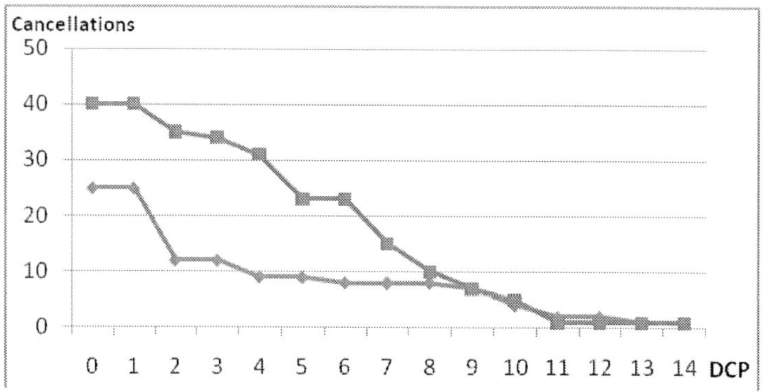

Figure 3. Two curves with similar patterns in the beginning but different cancellation numbers at DCP(0)

Another possible feature hidden in the curves is what we call recency effect in this study. In short, a cancellation curve which is close to the current date is regarded to be more influential than another which is away from the current date. This instance selection concept is also advised in the literature. For example, Shimodaira (1996) applied the concept of moving window data learning method to dynamically utilize samples over time. Chen and Dai (2005) designed a Circular Back-propagation Neural Network and used Discounted Least Squares (DLS) for parameter estimation. In their paper, the concept of DLS is to treat each sample with a time-based weight, which is a variant of the recency effect.

Since we aim to forecast the number of daily cancellations, it is also important to include the periodic effect of cancellation curves. As we can learn from the data, passengers in fact have different demand over each day-of-week. For example, we found higher cancellation numbers on Mondays and also weekends in the studied train service (see Figure 4). When a forecast target has very limited information in the beginning of the booking period, day-of-week will be a very significant variable for projecting future cancellations at DCP(0). However, as DCPs are approaching the service day, the role of day-of-week should be marginalized given more and more cancellation information.

2.2. Cancellation Patterns at Departure

In this part, we further analyze the distribution of cancellations at departure. Figure 5 shows the idea, and we can have the following observations. First of all, there seems no trend existed in the data. In order to verify the claim, we built a regression model utilizing a continuous trend index and six day-of-week dummies as independent variables. The result confirms the absence of trend and also the presence of day-of-week periodicity. Second, a permanent level shift is happened after point 225. Also after point 225, the variance of the studied series enlarges which implies the non-stationary property of the data. The shift of mean and the change of variance must be caused by some reasons; however, we have no any prior knowledge about the causes. In order to achieve model robustness, the model must be capable of knowing the change of data characteristics and adapting to a new environment. We will introduce the design of adaption in the model section.

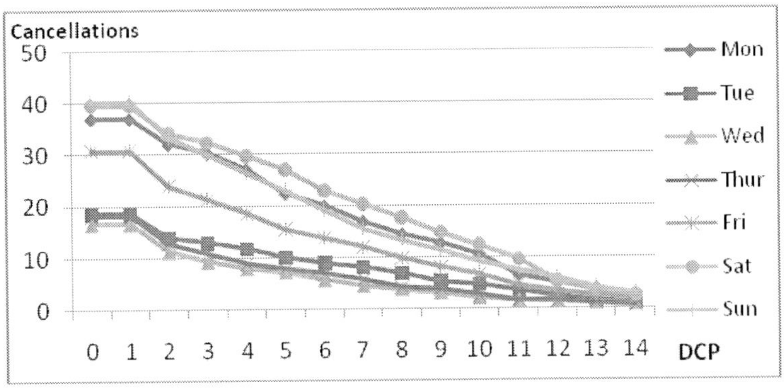

Figure 4. Average cancellation curves based on day-of-week

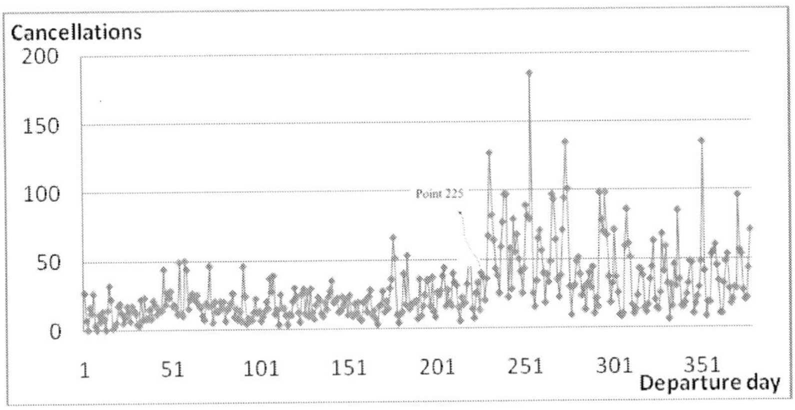

Figure 5. The number of cancellations at departure

3. MODELS

In the previous section, the reliability of information, recency effect, day-of-week periodicity, and adaption are regarded to be important characteristics in the studied cancellation data series. In this section, we will incorporate these four features into a case-based reasoning model. The proposal in fact has four sequential steps in order to generate the final cancellation forecast. We will introduce each step one after another as follows. In addition, we will also brief the concept of regression and pick up models with their variants.

3.1. Case-Based Predicting Model (CBP)

The proposed CBP has four sequential stages: similarity evaluation, sample selection, forecast generation, and parameter search.

3.1.1. Similarity evaluation

The first stage evaluates the similarity among available samples (*i*) in the database and a forecast target (*j*). Since cancellation information is updated at each DCP, the calculation needs to be dynamically updated at each DCP. In other words, we need to recalculate the similarity every time when new cancellation information becomes available, as shown in Equation (1) (*k=14~1*). In the equation, *k* represents the current DCP of the forecast target *j*; $x_{.,m}$ is the number of cancellations at DCP(*m*). Euclidean Distance is the most basic and also capable of evaluating the geometric difference between two curves. Based on this, the incorporation of temporal features seems to be helpful for upgrading retrieval performance. We will show how to achieve this goal in the following and verify the proposal in the empirical study.

First, we add the feature of information reliability on the base of Euclidean Distance. Here we assume that the contribution of information at each DCP follows an exponential distribution. The farther a DCP is away from the departure day, the less influence it gets. In order to have this effect, the reciprocal of the DCP index (*m*) is used as a weight in the equation. Furthermore, α, which is anticipated to be positive, is used as an exponent to seize the strength of information reliability in Equation (1).

The recency effect is reflected by computing the time difference between a sample *i* and the forecast target *j*. If time difference is large, then the sample *i* is expected to have a significant weight. On the other hand, a low weight will be applied to the sample *i* if time difference is small. A parameter (ω), which is also expected to be a positive value, is utilized to grab the effect of recency.

The periodic effect is incorporated by matching day-of-week between a sample *i* and the forecast target *j*. If day-of-week of *i* and *j* is the same, then ω = 0 in Equation (1); otherwise, ω = 1. Via this setting, if the sample *i* has a different day-of-week from the forecast target *j*, the sample *i* would have a significant penalty. The above mechanism seems to be reasonable for data retrieval. However, we want to emphasize on the point that the effect of day-of-week should be marginalized over DCPs since more and more cancellation information is accumulated. In the beginning of the booking period, day-of-week might be a good reference when the volume of cancellations is still low and unreliable to predict cancellations at DCP(0). When cancellation curves can show patterns over DCPs by themselves, the influence of day-of-week should become less important. In this study, we use the reciprocal of *k* which represents the current DCP to marginalize the effect of day-of-week. A parameter (γ), which is expected to be a positive parameter, is applied to seize the influence.

$$d_{j,i}^{k} = \sum_{m=k}^{14}\left(x_{j,m} - x_{i,m}\right)^2\left(\frac{1}{m}\right)^{\alpha}(j-i)^{\beta} + \gamma\left(\omega + \frac{1}{k}\right),\ k = 14\sim1 \qquad (1)$$

3.1.2. Sample selection

After the evaluation of similarity between all available samples and the forecast target, the next stage is to pick up a certain amount of similar samples for the following calculation. One important question to ask here is how many samples (indexed by *s*) should be chosen in order to obtain satisfactory predictive performance. In this study, the proposed procedure ranks all samples by similarity obtained in the first stage and tests three scenarios for the "best" selection of similar samples (*s=10, s=20, s=30*). The optimal size of instance selection is another important issue; nevertheless, we do not aim to find out this magic number in this study. Further investigation is necessary to answer the abovementioned question.

3.1.3. Prediction generation

Based on the selected samples with high similarity to the forecast target *j* from the previous stage, the next step is to project the final number of cancellations of the forecast target at DCP(0). The most naïve method is to calculate the simple average of cancellations in terms of all selected samples at DCP(0). However, the concept of average is not going to perform well since we have already indicated the importance of adaption in the data section. The prediction method here must be able to include novel (and also available) cancellation information. In order to achieve this effect, we have two strategies for generating final cancellation numbers.

First of all, the information of similarity obtained in the first stage is utilized to weigh each selected sample ($\frac{d_{j,n}^k}{\sum_{p=1}^{s} d_{j,p}^k}$). Samples with high similarity, without questions, obtain large influences. This strategy can be imagined as summing up all cancellation information up to the current DCP.

The second strategy emphasizes on the value of current cancellation information. We added a term ($\frac{x_{j,k}}{x_{n,k}}$) in the prediction equation so that current cancellation information in the forecast target (*j*) and corresponding cancellation information in the selected samples (*n*=*1*~*s*) can be compared. This ratio is utilized to modify the contribution of each selected sample.

One important point should be mentioned here. Both the previous two strategies should consider the location of the current DCP. If the current DCP is away from the departure day, we should use the strategies in a conservative way since information is unreliable in the beginning of the booking period. Once DCPs are approaching the departure day, we can adopt both strategies with high confidence. The reciprocal of the current DCP (*k*) is used as an exponent to represent the time-prospering confidence. Two individual parameters, δ and φ, are applied to seize the strength of the strategies, respectively. The whole formula is shown in Equation (2).

$$\hat{x}_{j,0}^k = \sum_{n=1}^{s} \left(\frac{\frac{1}{d_{j,n}^k}}{\sum_{p=1}^{s}\frac{1}{d_{j,p}^k}}\right)^{\left(\frac{1}{k}\right)^{\delta}} \left(\frac{x_{j,k}}{x_{n,k}}\right)^{\left(\frac{1}{k}\right)^{\varphi}} x_{n,0},\ k = 14{\sim}1 \qquad (2)$$

3.1.4. Parameter search

The last stage of the proposed model is to search for five applied parameters (α, β, γ, δ, φ). Instinctively, we can turn the whole forecasting system into a non-constrained optimization problem and apply appropriate algorithms for the search of parameters. Nevertheless, the proposed model is multi-stage and highly nonlinear. Deriving gradient information by using conventional gradient decent algorithms is expected to be difficult. Another possibility for parameter search is the use of metaheuristic methods such as genetic algorithm, particle swarm optimization, simulated annealing etc. However, one significant weakness of using these global optimization methods is the learning efficiency. For revenue management forecasting, efficiency is an important factor since frequent update and plenty products are usually coming together.

In this study, we applied Hooke-Jeeves algorithm to search the parameters (Himmelblau, 1972). The basic idea behind Hooke-Jeeves algorithm is the mix use of explanatory and pattern searches. The explanatory search explores the neighboring area for reaching feasible solutions. The design of the pattern search equips the algorithm with the capability to search in a possible direction (toward the optimum). One weakness of Hooke-Jeeves algorithm is the possibility of sticking into a local minimum. In order to improve the quality of the solution, a multi-start strategy was adopted in this study to test multiple initial seeds. We can now imagine the whole model as a two-tier procedure. The first tier is the proposed CBP model which is responsible for generating forecasts and also calculating performance indices such as mean square errors (MSE). The next tier is the searching mechanism which aims to find the best combination of parameters. The whole learning process is stopped if no further improvement of performance is attained; the procedure is shown in Figure 6.

Another key issue here is the use of samples. In this study, we have collected data and yielded 378 cancellation curves. We kept the latest 30 samples as the testing set. Another 30 samples which are dated just right before the testing set were taken as the validating set. The rest 318 curves were used as the training set. The model first uses the training and validating sets to determine the value of parameters. After that, the model applies the obtained parameters to compute forecasts and evaluate out-of-sample performance by using the testing set.

3.2. Regression Models

Regression is a convenient tool and can be used as a benchmark to predict cancellations in this study. The idea is to map the relationship between the cancellations on the departure day (DCP(0)) and a booking day (DCP(k)), as shown in Equation (3). Then this relationship can be applied to predict cancellations at DCP(0). (Equation (4)). Each DCP, as a consequence, has a corresponding regression model. It is also worth noting that regression only utilizes partial information of cancellation curves ($x_{i,k}$) in comparison with the proposed CBP model which uses all available cancellation information ($\sum_{m=k}^{14}(x_{j,m} - x_{i,m})^2$). However, including more cancellation variables in regression will result in misleading outcomes due to the problem of multicolinearity. We also investigated a variant of regression here which includes six day-of-week (dow) dummies, as shown in Equation (5). Equation (6) is used for the job of prediction.

$$x_{i,0} = \mu + \pi x_{i,k} + \varepsilon_i,\ k = 14\sim 1 \tag{3}$$

$$\hat{x}_{j,0} = \hat{\mu} + \hat{\pi} x_{j,k},\ k = 14\sim 1 \tag{4}$$

$$x_{i,0} = \mu + \sum_{h=1}^{6} \rho_h \mathrm{dow}_h + \pi x_{i,k} + \varepsilon_i,\ k = 14\sim 1 \tag{5}$$

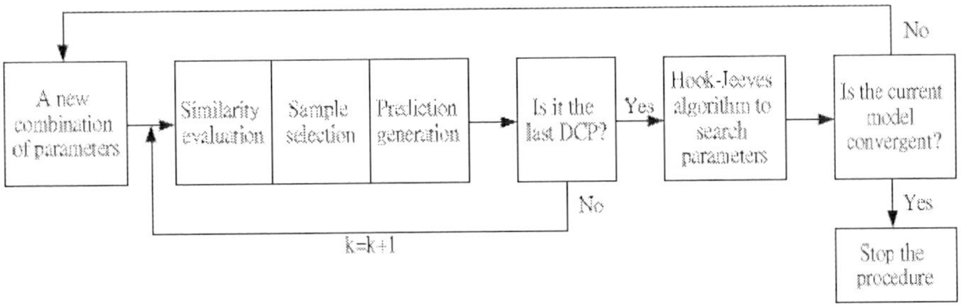

Figure 6. Learning procedure of the CBP model

$$\hat{x}_{j,0} = \hat{\mu} + \sum_{h=1}^{6} \hat{\rho}_h \, dow_h + \hat{\pi} \, x_{i,k}, \; k = 14 \sim 1 \tag{6}$$

3.3. Pick up Models

Instead of building a regression model, another simple alternative is to first calculate the average deviation of cancellations between DCP(0) and DCP(k) (Equation (7)). Then this average deviation is added onto the accumulated cancellation numbers based on the current DCP for the projection of final cancellation numbers (Equation (8)). In Equation (7) and (8), *L* is the number of samples used to calculate the average deviation. Analogously, the effect of day-of-week can also be incorporated in a pick up model. The data are initially divided into seven groups based on day-of-week. Then each group of data follows the procedure of Equation (7) and (8) and constructs respective pick up models.

$$\bar{d}_k = \frac{1}{L} \sum_{l=1}^{L} (x_{i,0} - x_{i,k}), \; k = 14 \sim 1 \tag{7}$$

$$\hat{x}_{j,0}^{k} = x_{j,k} + \bar{d}_k, \; k = 14 \sim 1 \tag{8}$$

4. Empirical Study

4.1. The Best Number of Selection

In the second stage of the proposed CBP model, the number of selected samples is left for determination. Three simple scenarios were tested here with *s=10, s=20,* and *s=30.* Mean square errors (MSE) given these three scenarios were calculated DCP by DCP, as shown in Figure 7. As we can learn from the figure, selecting too few samples (*s=10*) result in unsatisfactory performance. Increasing the number of selection to the next level can significantly improve predictive accuracy (*s=20*). When more samples are utilized, the improvement becomes marginal (*s=30*). More samples are not expected to have any more contributions based on this trend. As a result, we constantly select 30 similar samples for the following analysis and comparison. Another interesting point to mention is that the magnitude of improvement is apparently not constant. When a DCP is away from the departure day, selecting more samples has significant help in predicting future cancellations. Once the DCP is less than five, the predictive performance is almost the same regardless of the number of selection. This outcome suggests the use of different number of samples for different DCPs; however, finding the best way for sample usage is beyond the scope of this study.

4.2. Comparison with a Naïve CBP Variant

In order to show the effectiveness of the designs based on the extracted temporal features, i.e., information reliability, recency effect, day-of-week periodicity, and adaption, the comparison between the proposed temporal CBP model (TCBP) and a naïve CBP (NCBP) is investigated. It should be noted that NCBP is, in fact, a special case of TCBP. In Equation (2), if we assume $\alpha = \beta = \gamma = 0$, then we can obtain a purely Euclidean-distance based formula for similarity evaluation (Equation (9)). If we assume $\delta = 0$ and γ equals to a relatively large number, then we can have a weighted average formula for prediction generation (Equation (10)). The above two assumptions complete the settings of NCBP.

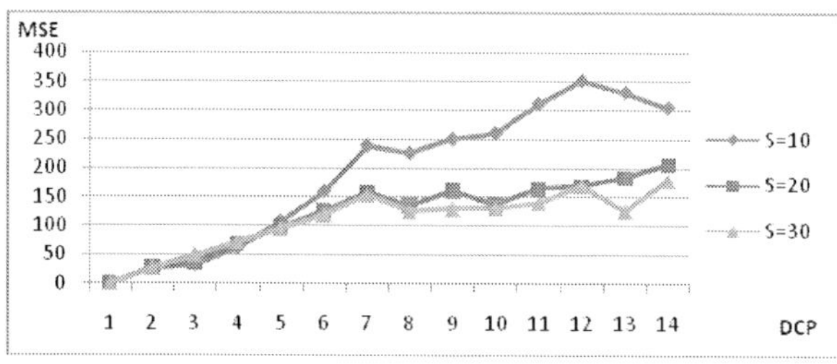

Figure 7. Comparison among three sample scenarios

Figure 8 first shows the best number of sample selection in NCBP. It is obvious to see that selecting more similar samples in the second stage in NCBP does not necessarily lead to improved performance. As a result, only ten samples are utilized in NCBP in this study. The comparison between TCBP (*s=30*) and NCBP (*s=10*) is shown in Figure 9. Apparently, NCBP cannot even compete with TCBP which reveals the value of including temporal features in the model design. In addition, the optimized value of parameters $(\alpha, \beta, \gamma, \delta, \varphi)$ in TCBP is (7.0, 1.93, 4.82, 0.03, 0.44). Comparing this parameter set with the value of parameters in NCBP, which is (0, 0, 0, 0, 999), also gives us a hint about the failure of NCBP.

$$d_{j,i}^{k} = \sum_{m=k}^{14} \left(x_{j,m} - x_{i,m}\right)^2, k = 14\sim 1 \quad (9)$$

$$\hat{x}_{j,0}^{k} = \sum_{n=1}^{s} \left(\frac{\frac{1}{d_{j,n}^{k}}}{\sum_{p=1}^{s} \frac{1}{d_{j,p}^{k}}}\right) x_{n,0}, k = 14\sim 1 \quad (10)$$

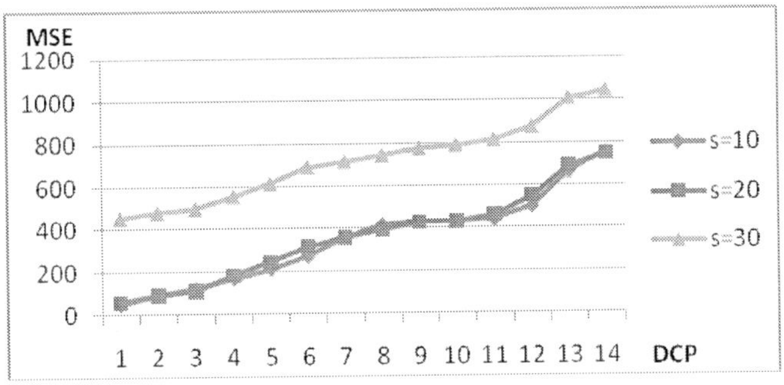

Figure 8. Comparison of the selected samples in NCBP

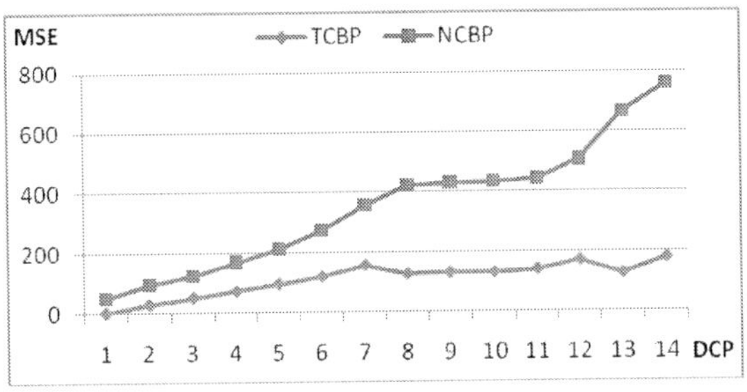

Figure 9. Comparison between TCBP (*s=30*) and NCBP (*s=10*)

4.3. Comparison with Four Benchmarks

In this section, performance comparison among TCBP and four introduced benchmarks, i.e., regression (Reg), regression with day-of-week dummies (Regdow), pick up (Pu), pick up with day-of-week classification (Pudow), is investigated (Figure 10). First of all, predictive accuracy improves (from DCP(14) to DCP(1)) once more cancellation information becomes available over DCPs, regardless of model prototypes. Second, the use of day-of-week dummies in conventional regression and pick up models results in the upgrade of predictive accuracy. Third, TCBP significantly outperforms all four benchmarks from DCP(14) to DCP(7). This outcome shows the potential of TCBP when cancellation information is limited in early DCPs. When DCPs are approaching the departure day, TCBP can still remain its competitiveness with four benchmarks.

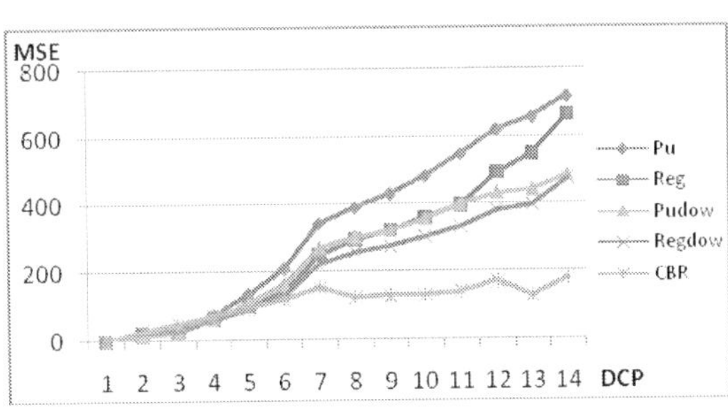

Figure 10. Comparison among TCBP and four benchmarks

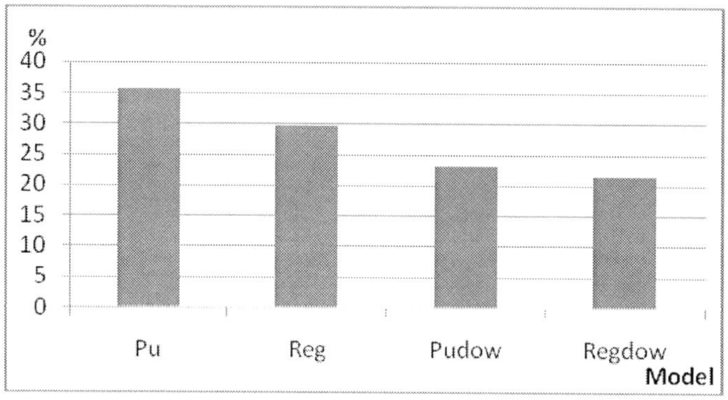

Figure 11. TCBP: percentage of improvement

We further calculate the percentage of overall improvement (regardless of DCPs) of TCBP in terms of four benchmarks (Figure 11). As we can learn from the figure, TCBP can have at least over 20% improvement of MSE over traditional benchmarks.

4.4. Distributions of the Estimated Parameters

In this section, we implement a scenario analysis to show the influence of parameters over DCPs. By doing so, we can check the assumptions we have made in the model section. We draw the distributions of five estimated parameters in Figure 12 (the order is from left to right and top to down). Except γ which represents the influence of recency effect and has nothing to do with DCPs, other four parameters have different values over DCPs. Here we use a scenario which will be described when needed in the following paragraphs to introduce the impact of each term. i.e., $(\frac{1}{m})^{7.0}, (j-i)^{1.93}, 4.82\left(\omega+\frac{1}{k}\right), (\frac{d^{k}_{j,n}}{\sum_{p=1}^{30} d^{k}_{j,p}})^{(\frac{1}{k})^{0.03}}, (\frac{x_{j,k}}{x_{n,k}})^{(\frac{1}{k})^{0.44}}$.

Assuming the current DCP is six, and as we can see from the upper left figure, the weights from DCP(14) to DCP(6) are very close. This phenomenon implies equal importance of information at early DCPs. Only DCP(1) has a large weight. As a result, the distribution of information reliability, instead of an exponential distribution, is more like a step function in the studied case.

If a sample is away from the forecast target in time, a significant weight will be multiplied to the current base. A high weight implies high distance and low similarity which decreases the probability of being selected. This outcome simply follows our assumption in the third section.

If day-of-week of a sample and that of the forecast target is not the same ($\omega = 1$) at DCP(6), a significant weight (on the red line) will be added to the current base. On the other hand, if day-of-week is the same, a trivial weight (on the blue line) adds almost nothing to the current base which increases the probability of being selected. In addition, as DCPs approach the departure day, both red and blue lines climb up which diminish the effect of day-of-week. This outcome also follows our assumption.

The similarity-based weight in the stage of prediction generation shows an almost-constant effect. Here we assume $\frac{d^{k}_{j,n}}{\sum_{p=1}^{s} d^{k}_{j,p}} = 0.5$ for the demonstration. As you can see from the right figure in the second layer, this ratio will be authentically reflected regardless of DCPs. Although we assumed the "gradual" importance of the similarity-based weight over DCPs in the model section, the outcome here indicates the "immediate" importance of the similarity-based weight at each DCP.

Last but not least, the bottom figure suggests us to be conservative in early DCPs while using the information from $\frac{x_{j,k}}{x_{n,k}}$ for improving adaption. Here we also assume $\frac{x_{j,k}}{x_{n,k}} = 0.5$ for the purpose of demonstration. As we can learn from the graph, although $\frac{x_{j,k}}{x_{n,k}}$ equals to 0.5 at DCP(6), TCBP multiplies 0.73, rather than 0.5, to the current base. This is, once again, because of unreliable cancellation information in early DCPs. Once DCPs approach the departure day, this adaptive ratio should be reflected in an authentic way (see the trend from DCP(5) and DCP(1)). This outcome follows our assumption in the model section.

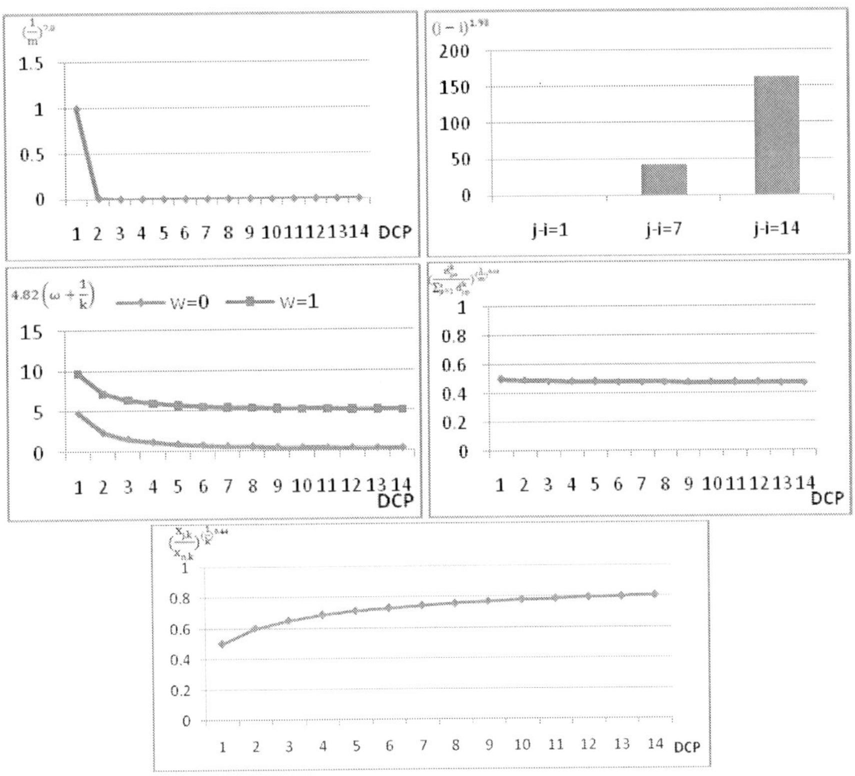

Figure 12. Distributions of the influential parameters (from left to right, top to down)

Conclusion

Accurate cancellation forecast is a key to successful allotment of perishable services (or products). Without accurate forecast, the company faces a challenge to oversell or spoil its inventory with high possibility. This study proposed a temporal case-based predicting model for cancellation forecasting. A four-stage procedure including similarity evaluation, sample selection, prediction generation, and parameter search is established. The comparison between the proposed TCBP and NCBP reveals the value of model design based on the extracted temporal features (i.e., information reliability, recency effect, day-of-week periodicity, and adaption). In addition, the competition among TCBP and the benchmarks shows the great potential of TCBP for the research problem. At least 20% improvement of MSE can be achieved by using TCBP.

It should be noted that all results reported in this study are based on a single case. In fact, we do not expect the proposed model to be an all-time winner in all data situations. On the other hand, we anticipate seeing regression and pick up models as winners in some cases. For example, when cancellation curves show highly linear relationships, regression and pick up models might perform very well. Two interesting extensions can be investigated following this line. First, more empirical studies based on heterogeneous data are needed to further confirm the validity of the proposed model. Data can come from multiple industries with the behavior of cancellation such as hotels, restaurants, rental cars, golf courses, airlines etc. More importantly, we can try to identify the situations for using the proposed technique versus conventional methods.

Since the proposed model is flexible and does not have a unique form (either additive or multiplicative), other variants based on the same spirit are also possible to improve predictive accuracy comparing with benchmarks. In short, we can define various formats of each influential term and then turn the forecasting system into a combinational optimization problem.

ACKNOWLEDGMENT

The author would like to show his gratitude to Prof. Chien-Hung Wei at the Department of Transportation and Communication Management Science in National Cheng Kung University in Taiwan for his great support.

REFERENCES

Chen, S. & Dai, Q. (2005). Discounted least squares-improved circular back-propagation neural networks with applications in time series prediction. Neural Computing & Applications, 14(3), 250-255.

Garrow, L. A. & Koppelman, F. S. (2004a). Multinomial and nested logit models of airline passengers' no-show and standby behaviour. *Journal of Revenue and Pricing Management, 3(3)*, 237-253.

Garrow, L. A. & Koppelman, F. S. (2004b). Predicting air travelers' no-show and standby behavior using passenger and directional itinerary information. *Journal of Air Transport Management, 10(6)*, 401-411.

Gorin, T., Brunger, W. G. & White, M. M. (2006). No-show forecasting: A blended cost-based, PNR-adjusted approach. *Journal of Revenue and Pricing Management, 5(3)*, 188-206.

Himmelblau, D. M. (1972). Applied Nonlinear Programming. U. S. A.: McGraw-Hill.

Hueglin, C. & Vannotti, F. (2001). Data mining techniques to improve forecast accuracy in airline business. In F. Provost, & R. Srikant, (Eds.), *Proceedings of the Seventh ACM SIGKDD International Conference on Knowledge Discovery and Data Mining*, (438-442). San Francisco, U. S. A.

Kimes, S. E. (2005). Restaurant revenue management: Could it work?. *Journal of Revenue and Pricing Management, 4(1)*, 95-97.

Lawrence, R. D., Hong, S. J. & Cherrier, J. (2003). Passenger-based predictive modeling of airline no-show rates. In P., Domingos, C., Faloutsos, T. Senator, & L. Getoor, (Eds.), *Proceedings of the Ninth ACM SIGKDD International Conference on Knowledge Discovery and Data Mining*, (397-406). Washington D. C., U. S. A.

Lee, A. O. (1990). Airline reservations forecasting: probabilistic and statistical models of the booking process. U. S. A.: MIT Press.

Lemke, C., Riedel, S. & Gabrys, B. (2009). Dynamic combination of forecasts generated by diversification procedures applied to forecasting of airline cancellations, *In IEEE Symposium on Computational Intelligence for Financial Engineering Proceeding*, (85-91). Nashville, U. S. A.

Makridakis, S. & Hibon, M. (2000). The M3-competition: results, conclusions, and implications. *International Journal of Forecasting, 16(4)*, 451-476.

Neuling, R., Riedel, S. & Kalka, K. (2004). New approaches to origin and destination and no-show forecasting: Excavating the passenger name records treasure. *Journal of Revenue and Pricing Management, 3(1)*, 62-72.

Ridel, S. & Gabrys, B. (2007). Combination of multi level forecasts. *Journal of VLSI Signal Processing, 49(2)*, 265-280.

Romero Morales, D. & Wang, J. (2008). Passenger name record data mining based cancellation forecasting for revenue management [http://www.optimization-online.org/DB_FILE/2008/04/1953.pdf].

Schwartz, Z. & Hiemstra, S. (1997). Improving the accuracy of hotel reservations forecasting: curves similarity approach. *Journal of Travel Research, 36(1)*, 3-14.

Shimodaira, H. (1996). A method for selecting similar learning data in the prediction of time series using neural networks. *Expert Systems with Applications, 10(3/4)*, 429-434.

Smith, B. C., Leimkuhler, J. F. & Darrow, R. M. (1992). Yield management at American airlines. *Interfaces, 22(1)*, 8-31.

Tsai, T. H. (2009). A temporal case retrieval model to predict railway passenger arrivals. *Expert Systems with Applications, 36(5)*, 8876-8882.

Weatherford, L. R. & Kimes, S. E. (2003). A comparison of forecasting methods for hotel revenue management. *International Journal of Forecasting, 19(3)*, 401-415.

Wickham, R. R. (1995). *Evaluation of forecasting techniques for short-term demand of air transportation*, U. S. A.: MIT Press.

Wu, K. T. & Lin, F. C. (1999). Forecasting airline seat show rates with neural networks. In *IEEE Neural Networks Council & International Neural Network Society* (Eds.), International Joint Conference on Neural Networks (3947-3977). Washington D. C., U. S. A.: Institute of Electrical and Electronics Engineers.

In: Case-Based Reasoning: Processes Suitability...
Editors: Antonia M. Leeland
ISBN: 978-1-61728-352-9

Chapter 5

PROVISION OF SAFETY FOR TECHNOLOGICAL SYSTEMS WITH THE AID OF CASE-BASED REASONING

A.F. Berman, O.A. Nikolaychuk, A.I. Pavlov and A.Yu. Yurin *

Institute for Systems Dynamics and Control Theory, Siberian Branch,
Russian Academy of Sciences, Irkutsk, Russia
Lermontov st., 134, nikoly@icc.ru

ABSTRACT

The problem statement, and the process of solving the problem, bound up with the investigation of safety related to complex technological systems on the basis of the method of case-based reasoning is considered. Case-based expert systems provide decision support on the principle of analogy and on the basis of experience data available.

Conceptualization and formalization of data and knowledge, which represent the state of technological systems, have been conducted. Principal properties and hazardous states of the technological systems have been identified. Hazardous states are represented in the form of a cause-effect sequence of states that are characterized by the increased risk level: failure, accident, emergency, and technogenic catastrophe. An object-oriented model of a case, which describes the proposed dynamics of hazardous states of some technological object on account of the complex technological system's structure and the hierarchy of hazardous states, has been developed. The elaborated case-base contains information about 200 incidents and failures, which have taken place at the USSR and Russian chemical and petrochemical enterprises.

The algorithms and the software system, which provide for finding the solution in the hierarchical space of cases, as well as for adaptation of the solution on the basis of production rules and analytical models, have been elaborated.

At the end of elevating the efficiency of the case retrieval procedure we have conducted the indexing of cases by using elements of the object-oriented model. The

* Corresponding author: *thtsai@mail.ncku.edu.tw*, +886-6-2757575 ext. 53270 ext. 5041.

software system provides for decision support in the processes of providing for the safety of technological systems and compensating for the consequences of failure, while including the solving of problems related to identifying hazardous states, determining their causes, assessing the degree of hazard and forecasting scenarios of the evolution of hazardous states.

A real example, demonstrating the decision support in the process of determination of the causes of failure, which is provided for by the software proposed, is considered.

1. INTRODUCTION

The safety of technological objects is one of the investigation problems characterized by priority character. Its solution presumes solving a set of problems, which necessitates competence in many scientific disciplines [1-4]. Thus, the problem related to the provision of safety of technological objects is multidisciplinary.

The application of the methods and aids of artificial intelligence (AI) allows one to substantially elevate the efficiency and the quality of decisions made under fuzzy conditions. This is especially true in case of complex and insufficiently formalized problems, such as the problem related to the safety of technological objects [5-14]. In particular, the method of *expert systems* (from now on – ESs) provides for the application of the knowledge of qualified specialists useful in solving complex problems, which are characterized by a high degree of uncertainty.

ESs allow the researcher to obtain an acceptable decision within a short time interval without deep knowledge about the scrutinized events or processes. The knowledge of specialists is represented in such an ES in the form of some formalized description; for example, in the form of (i) some sequence outlining the progress of some events and (ii) some decisions made in the process of the investigation or event control for provision of safety or at least for minimizing the losses. Operation of such an ES is substantially dependent on the quality of the formalization of the knowledge, which may be represented in the form of production rules, frames, etc., and also in the case-based form [15-18].

The case-based approach is used in solving such problems, for which there are descriptions of their decision. Some examples of the application of the case-based approach in the investigations, which are related to the object's technical state, and some examples of defining the causes (reasons) of critical failures of technological systems, can be found in [5, 6, 9, 10], and as far as investigation and provision of technogenic safety is concerned, – in [8]. An expert system, which is based on the case-based approach and is oriented to prevent failures in the design stage, can be found in [11]. In [12, 13], there are described intelligent systems intended for the support of failure analysis. The authors used an interesting combination of the case-based approach, genetic algorithms and neural networks.

Many experts have come to the conclusion that the improvement of the quality, and the high efficiency of the decisions made at the end of the provision of safety for complex technological systems, on the basis of methods and aids of artificial intelligence, represent a very important issue, which necessitates elaboration of the respective methodological grounds, algorithms (including mathematically represented algorithms) and intelligent software. All these components are essential for successfully solving the problem of safety provision.

The objective of the present work utilizes the application of the case-based approach for the purpose of solving the problems related to the provision of technological systems' safety. We employ the technological systems designed for the needs of chemical and petrochemical industries. In this connection, we have conducted the following creative work:

- conceptualization of the respective data and knowledge;
- design of the case model;
- elaboration (and grounding) of efficient techniques of obtaining relevant information;
- development of the algorithm and a software, whose intention implies the provision of safety of the technological objects.

2. Conceptualization of Data and Knowledge

Any Technological Complex (TCom) possesses a hierarchical structure (Figure 1) and is described by a set of properties, which characterize it during the product life. In the process of its functioning, such a technological system may exist in one of the following two types of technical states: operable and inoperable (nonserviceable). In its turn, the inoperable state may manifest itself as either a safe state or a hazardous state. Some limit states (with respect to safety criteria) may represent a boundary between them.

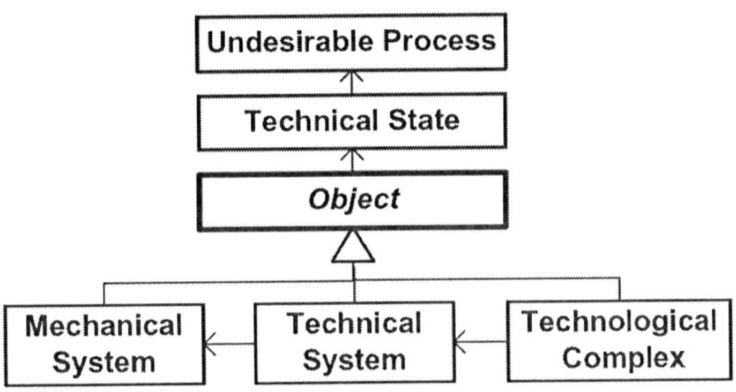

Figure 1. The hierarchical structure of Technological Complex.

Hazardous states may be classified into the following stages: the stage of Failure (the F-stage), the stage of Accident (the A-stage), the stage of Emergency (the E-stage), and the stage of Technogenic Catastrophe (the TC-stage) (Figure 2). A-stage is considered to be the initial stage in the development of E-stage and TC-stage. Each of the stages is described in terms of a set of events. Each event is described with the use of a set of parameters; their values and the functional dependences characteristic of their variations. The descriptions of events may repeat at different moments in time, and may differ in the values of parameters.

The F-stage is characterized by parameters of failure, e.g. effects of such a failure or the frequency of such failures. The A-stage is characterized by the parameters defining the state of the technological system, which exceed the permissible values of parameters, but – in some cases – may be brought to norm; otherwise, their further variation may be terminated. A-stage provides for a time reserve for organizing preventive actions, and it is provided by safety factors. If the stage has a sufficient duration, then there is a possibility of its early identification, and the ability to make decisions related to the actions needed to prevent development of E- and TC-stages. Decomposition of the process related to the formation of Technogenic Catastrophe into several stages allows one to conduct a deep analysis of the scenarios, events and states, which precede and accompany A-stage, E-stage and TC-stage [3, 4, 7]. In some cases, the A-stage elapses very quickly or instantly, which is conditioned by the imperfect character of the technological process control system and/or by designs inefficient in the aspect of provision of safety, which fail to provide for some time reserve needed for the decision making related to preventing the development of A-stage.

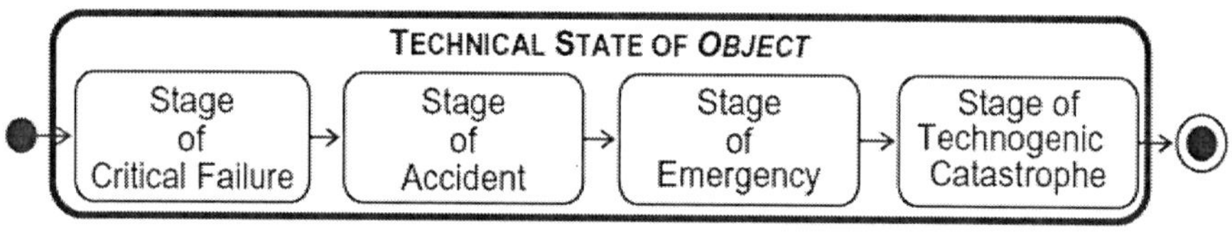

Figure 2. Stages of hazardous state.

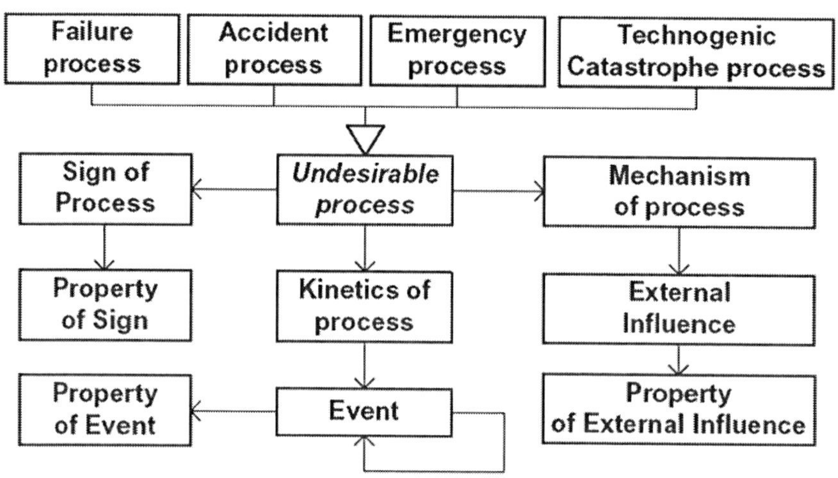

Figure 3. The object-oriented model of undesirable processes for technological system.

Proceeding from classification of hazardous states into stages, which differ in the degree of hazard, and assuming that there may be some safety changes in virtue of undesirable processes in the system in the process of its interaction with the environment, we are ready to consider the object model of these processes in UML (Unified Modeling Language) (Figure 3) [19].

Each process is described via explication of its mechanisms, its kinetics and its signs [2]. The mechanism of an undesirable (hazardous) process is characterized by the set factors, which influence the technological system. The kinetics and the dynamics of the undesirable (hazardous) process reflect the reaction of the technological system to a given influence in time in the form of events. The signs of its manifestation characterize the effects (consequences) conditioned by the external influences directed to the system.

3. The Case-based Approach

The database containing the information related to the failures of the equipment empoyed by Russian enterprises of petrochemistry [20], i.e. the information about the causes of cases, which presume failures and accidents, has formed the initial basis for the idea of automation and informatization of the processes at the end of solving the problems bound up with the provision of technological systems' safety on the basis of the method of case-based reasoning [17].

The model of a case. The model of a case represents some compact description of knowledge about some problem situation and its solution, and contains the following two principal components: the component containing the description of the problem and the component of decision making related to this problem. Consider the components of this case-based model:

$$Case = \langle \mathrm{Problem}, \mathrm{Decision} \rangle \quad (1)$$

The component of the problem's description includes the object's properties, which reflect its states on different stages of manufacture and exploitation. These properties have been obtained through the process of investigation of the cases (Accidents and Emergencies), and, in this connection, the assessment of the object's state. In our case, the description of the problem is based on the aggregated model of the scrutinized object, which can be found in [21].

$$\mathrm{Problem} = \langle O^{Stat}, O^{Dyn} \rangle,$$

where O^{Stat} – static information model of the object (which includes information representing the object on the design stage); O^{Dyn} – dynamic information model of the object (which includes information representing the object on the stage of its exploitation).

The static model (on the stage of design):

$$O^{Stat} = \langle EI^D, P^D \rangle,$$

where EI^D – presumed factors of external influences; P^D – presumed object's properties: $P^D = \langle N_U, TechChar, TechR, TechC, P_S, P_R, TechState^D \rangle$, N_U – number of unique elements inside the object; $TechCha$ – technical characteristics; $TechR$ – technical requirements; $TechC$ – technical conditions; P_S – properties of the system of safety, for example, availability and types of protective devices; P_R – properties of the system of reliability, for example, availability and the kinds of embedded diagnostic systems; the level of confirmation of periodicity of technical state assessment; $TechState^D$ – object's initial technical state, $TechState^D = \{H_i\}$, H_i – i-mounting heredity, D – index of design stage.

The dynamic model (on the stage of exploitation):

$$O^{Dynam} = \langle T, IF^E, P^E \rangle,$$

where T – discrete time moments of the object's life cycle; IF^E – influencing factors related to the object's exploitation; P^E – exploitation properties of the object (this set corresponds to the structure of properties related to design); $TechState^E$ – exploitation states of the object, $TechState^E = \{UP_l\}$, UP_l – l-undesirable processes for the object; $TechState^E = \{TechState^k, TechState^k \to TechState^{k+1}\}$, $\to$ – the cause-effect relation; k – is the index used for the stage of undesirable processes (failure, incident, accident, technogenic catastrophe), $k = \overline{1,4}$. At the moment of beginning of functioning of object T_0, its design and its exploitation states coincide. This fact may be expressed by the following identity $TechState^D \equiv TechState^E$.

$$UP_l = \{UP_{lk} | UP_{lk} = \langle M_{lk}, K_{lk}, S_{lk} \rangle,\ M_{lk} = \langle EI^E, P^E \rangle,\ UP_{lk} \to UP_{lk+1},\ M_{lk} \to \{K_{lk}^{mt}\}_m,$$
$$\{K_{lk}^{mt}\}_m \to \{S_{lk}^{pt}\}_p\}$$

where UP_{lk} is the k-stage of the l-undesirable process; M_{lk} is the mechanism of the k-stage of the l- undesirable process (the set of object's properties and the set of factors, which influence the object); K_{lk} is the kinetics of the k-stage of the l- undesirable process (the set of events, which describe hazardous processes and phenomena); S_{lk} are the signs of the k-stage of the l-undesirable process; K_{lk}^{mt} is the m-event of the k-stage of the l-undesirable process at the time moment t; S_{lk}^{pt} is the p-signs of the k-stage of the l- undesirable process at the time moment t; l is the index of the undesirable process; m is the index of kinetics of the undesirable process; p is the index of the indicator of kinetics of the undesirable process; $\to$ is the sign of cause-effect relations.

The component of the case, which describes the decision related to the problem, includes the following elements:

$$Decision = \langle Cause, Controling\ actions \rangle,$$

where *Cause* is a description of the total complex of interconnected causes, which may lead (the stage of design), or have already led (the stage of exploitation), to some hazardous states; *Controlling actions* represent a description of the sequence of some controlling decisions, which have been made at the end of preventing (the stage of design) or localization, liquidation and minimization of negative effects of A-stage, E-stage or TC-stage (the stage of exploitation).

Consider the object-oriented model of case on the basis of (1) and the object-oriented model of concepts "incident" [6] and "accident," where the concept of "incident" implies damage of a system, violation of some parameters and other deviations from the norm, which may lead to A-stage, E-stage or even to TC-stage.

The case represented in Figure 4 describes a scenario of an Accident. It presumes the following information is available: the name of the object and its hierarchical belonging (technological complex – technical system – mechanical system); actual exploitation conditions; observed parameters related to variation of the technical state and their external manifestation; a sequence of states, which have led to each hazardous stage, while including the properties of these states; causes of states; the structural genesis, i.e. the constructive element, which has become the cause of the hazardous state; the organization genesis, i.e. violations and imperfections, which have been admitted on all the stages of its life cycle; the decision made and related to control of the undesirable state, effects (consequences) of undesirable processes, etc.

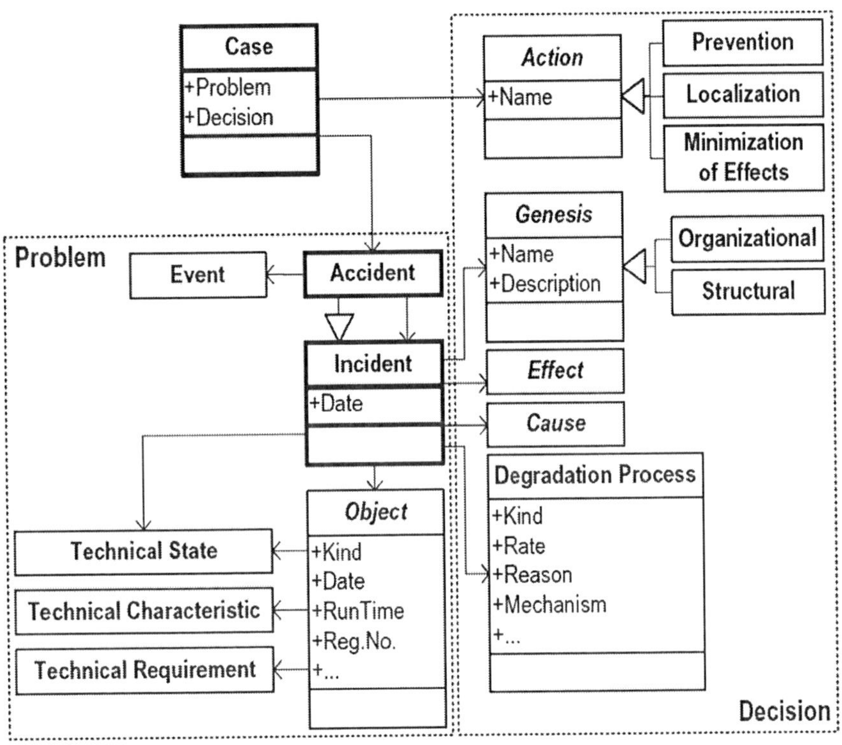

Figure 4. The fragment of case model.

Since each element of the hierarchical structure of a technological complex (TCom) exists in some state, the relation «part of» between the elements of the structure conditions the cause-effect relations between their technical states: the technical system is a component of TCom and simultaneously is the cause of occurrences of undesirable stares of the TCom. In this connection, it is possible to speak about the existence of the cause-effect hierarchy of the cases: the case of a technical system describes the cause of the undesirable state of the TCom (Figure 5):

$$Space\_of\_cases = \sum_{ij} Case_{ij},$$

where i – index of the hierarchy of TCom, j – index of the hierarchy of technical systems.

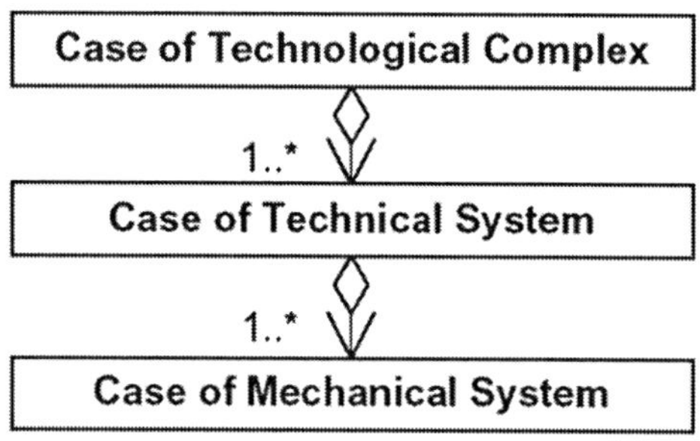

Figure 5. The structure of space of case.

Case retrieval. Implementation of the case-based approach is conducted by forming a case base, which is represented according to an elaborated model, and by organizing a procedure oriented to the analogs retrieval in this base. Such an analog presumes the case, whose description corresponds (to the largest possible degree) to the description of the problem situation under scrutiny, i.e. it contains the largest number of similar properties in comparison to other cases.

The process of the *search* and *retrieval* of analogs employs elements of the theory of *pattern recognition*. Hence, a pattern of the object to be recognized, which corresponds to the concept of "*case,*" is formed.

In order to provide for a successful *retrieval,* the system of recognition was developed. In particular, we have made a dictionary of parameters and described the patterns with the aid of this dictionary.

The dictionary of parameters has a complex structure. It has been subdivided into the following groups:

- the properties describing the *structural belonging* of the object have a hierarchical structure of the type: type → subtype;
- the parameters, describing the technical state labeled "failure", have a hierarchical structure of the type: failure→ external manifestation of the failure.
- the parameters, which describe such technical states as "*incident*", "*accident*" and "an technogenic catastrophe" have a hierarchical structure of the type: scenario → event → properties of event.

This structure allows one to apply elements of *the procedure of sequential solutions* for the purpose of solving the problem of retrieval [22]. The process of obtaining a solution is conducted sequentially, step-by-step, while considering separate fragments (groups of parameters) of the case. The process of search utilizes step-by-step *lowering the dimension* of the space of parameters and "reducing" the space of cases into a hyper-sphere or a hyper-cube. Furthermore, the domain of the hyper-cube includes the patterns, which have a description, which is *in some sense similar* to the description considered in terms of an analyzed set of criteria.

For the purpose of elevating the efficiency of the procedure related to retrieval (reducing its computational capacity) we have conducted the indexing of the cases on the basis of a dictionary of parameters, which we have developed. Normally, an individual set of parameters, which describe its properties (structural belonging of the object; its current technical condition, etc.) is put in correspondence to each of the cases.

Depending on the complexity of the description (availability of a hierarchy of properties and the type of parameters: either determined or logical ones), the indices (descriptors) represent either binary sequences (…01001…), or sets of corteges ({…, P_i , …}, $P_i = \langle n, v, i, c \rangle$, where n is the name of property; v is its value; i is the importance or the informative weight of the property; c is the constraint imposed on the interval of values). The constraint c defines the interval of values, within the frames of which the value of the property may characterize the value related to the similarity assessment: in case of occurrence of the value of some property beyond this interval, the similarity is absent, and the value of the similarity assessment is zero.

The procedure of case retrieval presumes (i) exhaustive search of variants and (ii) comparing of sets of the indices (related to the scrutinized cases), which correspond to a definite state. The result of such a search is represented by a set of analogs, which are ordered according to the similarity assessment (closeness) of the descriptions [the degree of uncertainty (incompleteness) of similarity estimates obtained is also indicated]. The similarity assessment of the descriptions is formed as a result of comparing descriptions of the cases to the end of revealing qualitative or quantitative differences [6, 8]. Qualitative differences reflect the absence of some parameters in the description of cases, while quantitative differences reflect the presence of different values which can be attributed to the same parameters. The level of similarity of cases (closeness in the description of cases) is computed as a distance between the cases in the space of parameters. It is measured by Minkovski metric [18], which is a generalization of the metric of city districts (which is used in processing binary vectors) and the Euclidean distance, which is used in processing sets of corteges.

$$dist_{Minkowski}(\bar{x}, \bar{y}) = \left(\sum_{i=1}^{n} |x_i - y_i|^p\right)^{1/p}$$

Parameter p defines whether it behaves like the metrics of city districts ($p=1$) or like the Euclidean distance ($p=2$).

When computing the similarity assessment, we take account of the expert's personal preferences in the form of the estimate of the importance of some or another parameter in the description of a definite system's state. The importance of such parameters is determined either by the user (who uses the linguistic scale) or as a result of the analysis of the case base. It corresponds with the frequency of the occurrence of the parameter in the description of problematic situations on account of a hierarchy of indicators described by the object-oriented model.

Besides the value of the uncertainty estimate, this stage presumes the computation of the degree of uncertainty for the estimate of similarity [23]. The definiteness of the assessment of the similarity is substantially influenced by the values of parameters, which the user has failed to determine (since the user does not possess some sufficient level of knowledge, skills and aids, which are needed for the purpose of defining this value); furthermore, the parameter, which has the largest importance, possesses some greater influence on the uncertainty of the closeness estimate. A situation, where an analog with a high estimate of similarity has a high estimate of uncertainty of the decision, is quite possible. The assessment of the uncertainty of a decision at a given stage (at a give level of development of the algorithm) this high does not influence the value of the similarity estimate; the situation shows the expert the implausibility of the result obtained, its inexactness, its fuzziness, while granting the expert the possibility of taking the measures related to redefining the initial data bound up with the object's description. The decision to redefine the initial data is normally made on the basis of recommendations given by the software, which are related to the selection of the methods and aids for redefining the values of parameters.

In the case of *reuse* of the decisions made in some *similar problem situations*, there appears the need in *adaptation of the decision* (earlier made for a similar problematic situation) to the the problematic situation under scrutiny on account of its new conditions and peculiar properties. In this case, it is possible to use the so called *transformational adaptation* [17, 24].

As far as the given type of adptation is concerned, the decision related to a *new* problematic situation is formed by *copying the elements of the decision made* in an earlier similar problematic situation and, next, by *transforming the previous decision*. The advantage of the transformational adaptation is that it requires the description of only the problem situation; unlike the case of generative adaptation, which necessitates the description of the whole trace (solving process) for efficient application; which, in turn, requires detailed formalization of the problem domain. In the present stage of our investigations, it is quite difficult to ensure a sufficient level of problem domain formalization to the end of application of generative adaptation. So, the application of such an approach is the matter of further investigations.

In our investigation, *adaptation* of the decision is conducted by the user. The decision making process presumes the following *two principal forms*: i) transformation of the *description* and/or ii) transformation of the *decision* for the problematic situation under scrutiny.

The process of transformation related to the *description* of the problematic situation under scrutiny is conducted by repeated concretization (qualitative redefining the description of the case) and (or) by redefining the values of the parameters, what leads one to a repeated search for new analogs and new decisions.

The process of redefining the values of parameters may be realized on the basis of analytical formulas or expert knowledge. The analytical formulas describe some aspects of the dynamics related to a accident or a TC. For example, consider the description of i) the speed of spread of fire, ii) the power of the shock wave, iii) the discharge of the fluid, which is running out, etc. The expert knowledge, which is normally represented in the form of production rules, reflect the mechanism, the kinetics and the signs of the undesirable processes, which take place or may take place for a technical system.

Transformation of the *decision* for the problem situation is executed by changing the elements of the decision. The latter is realized also on the basis of i) expert knowledge about the processes and the phenomena represented in the form of respective problem domain models and ii) the usage of production rules.

The principal algorithm. The algorithm of the investigation constructed on the basis of the case-based approach includes:

- designation of a hazardous object under design or choosing an object, on which an undesirable event is observed;
- intention (or description) of some external manifestation of undesirable processes or a set of events, which are observed for the object;
- description of values of the parameters characterizing each of the events;
- choosing the problem to be solved: identification of an accident or forecasting of an accident, or else finding the genesis of an accident;
- retrieval of cases (analogous);
- selection (by the user) the most similar case;
- adaptation of the most similar case to the current case.

In the course of solving the problems related to identification or forecasting (prediction) an accident, it is necessary to organize the retrieval of cases, which contain a description of the events indicated. As a result, it is possible to propose a list of cases, which contains descriptions of possible scenarios of the development of an accident; furthermore, the scenarios may include the events, which, for some reason, have not been indicated by the user, but represent an important part of the undesirable process. When necessary, it is possible to construct an event tree intended for the risk analysis of a current state for the technical system. The event tree may also include the forecasted events for the next level in the hierarchy of the scrutinized object, i.e. for a technological complex. Next, on the basis of analysis of the cases obtained, it is possible to redefine the parameters of the events already described and add the information about the events from the most similar cases. Additional information gives the possibility to conduct a new search and to propose a more exact (correct) set of scenarios.

In the course of solving the problems related to identifying or forecasting an accident in the process of exploitation, it is necessary to form (plan) a system of actions necessary for localizing and minimizing any negative effects (consequences) of the accident for each event of the scenarios,on the basis of operations on the event tree.

In the course of solving the problems related to forecasting accidents on the stage of design, it is necessary to form a system of actions (a plan of actions) related to preventing any undesirable events and minimizing any negative effects (consequences), if any.

In the course of solving the problem related to the genesis, it is necessary to organize the search for the cases which contain the description of the events indicated. Next, it is necessary to choose the case, which is the most similar and which presents a description of the cause-effect complex related to formation of an accident (A), an emergency (E) or a technogenic catastrophe (TC), which represents an analog of scenario under scrutiny. Generalization of the information obtained may be organized as event tree of A, E and TC. The result of solving this problem is a system of actions related to preventing any undesirable events, which form the current scenario.

4. Implementation of the Software

The approach proposed has been implemented in the form of an intelligent software, whose set of program components (modules) includes: a hierarchical case base; case retrieval module; rule base (needed for the process of adaptation); a module that implements the process of inference according to the rules (the interpreter of the rules); a library of mathematical modules; and a library of wizards (Figure6).

In the capacity of the software for storage of cases we have used DBMS Cache [25], which provides for efficient representation of cases in the form of "objects" or "relation tables," which allows us to use the object-oriented representation of cases without any changes in the level of data storage.

The graphic user interface and the algorithms have been implemented in Borland Delphi [26]. In the process of implementation of the module for adaptation of the cases, we used CLIPS intended for constructing expert systems [27]. CLIPS allows one to organize the process of adaptation on the basis of deductive reasoning.

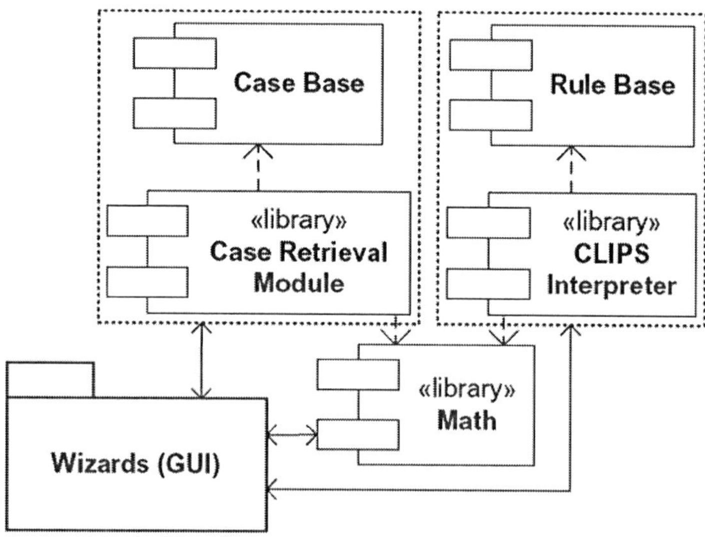

Figure 6. The Architecture of intelligent software.

The intelligent software includes a set of wizards, each realizing an algorithm of solving one of the problems and granting a user-friendly instructing interface efficient and comfortable for the user.

The intelligent software elaborated executes the following functions:

- registering (description) of the undesirable state (failure or accident) on the basis of the information available;
- discovering the cause-effect complex of the factors, which have conditioned the failure

(accident);

- planning the works related to redefining the causes of the case (when necessary);
-grounding necessary actions related to: preventing the causes of cases; provision and restoration the system's operability; minimization of the hazard and reduction (minimization) of the economic, ecological and social effects (consequences) and losses;
- identifying the undesirable state (failure or accident) and its registration into the case base;
- constructing the event tree on the basis of the information related to failures (accidents);
- generating a special report related to the failure (accident).

5. Example of Application of the Software

Consider a real example of Emergency and the possibility of making a decision with the aid of our intelligent system. Brittle failure of the girth weld on the pipeline exploited for the supply of synthesis gas into the column intended for synthesis of ammonium has initiated an accident, which provoked fire, a replacement of the synthesis column and caused substantial economic loss [28]. The scenario of the emergency may be described in the form of a sequence of events for each stage of development of a hazardous state (Figure 7).

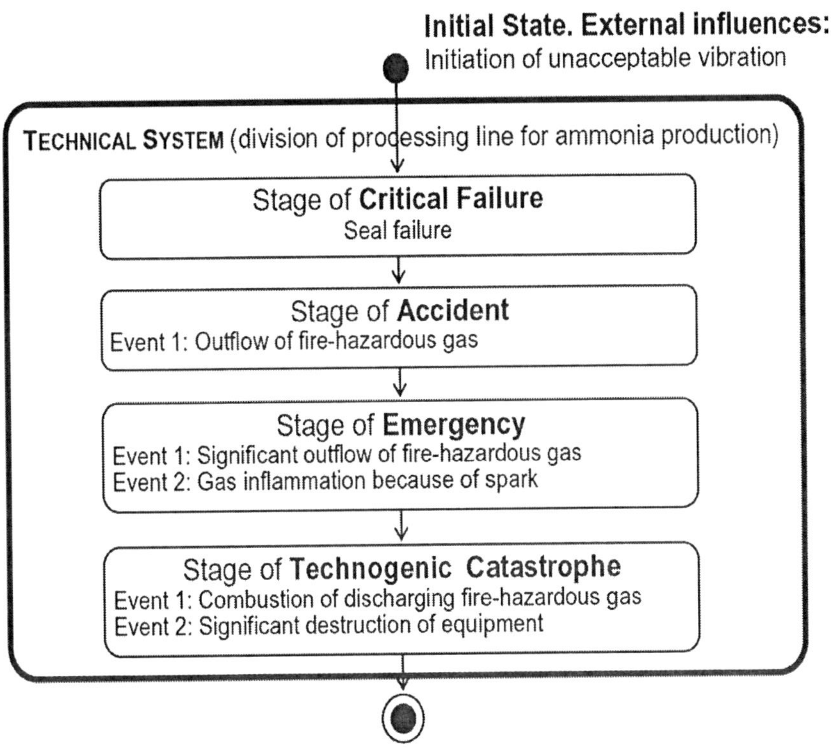

Figure 7. The example of description of the scenario of the technogenic catastrophe.

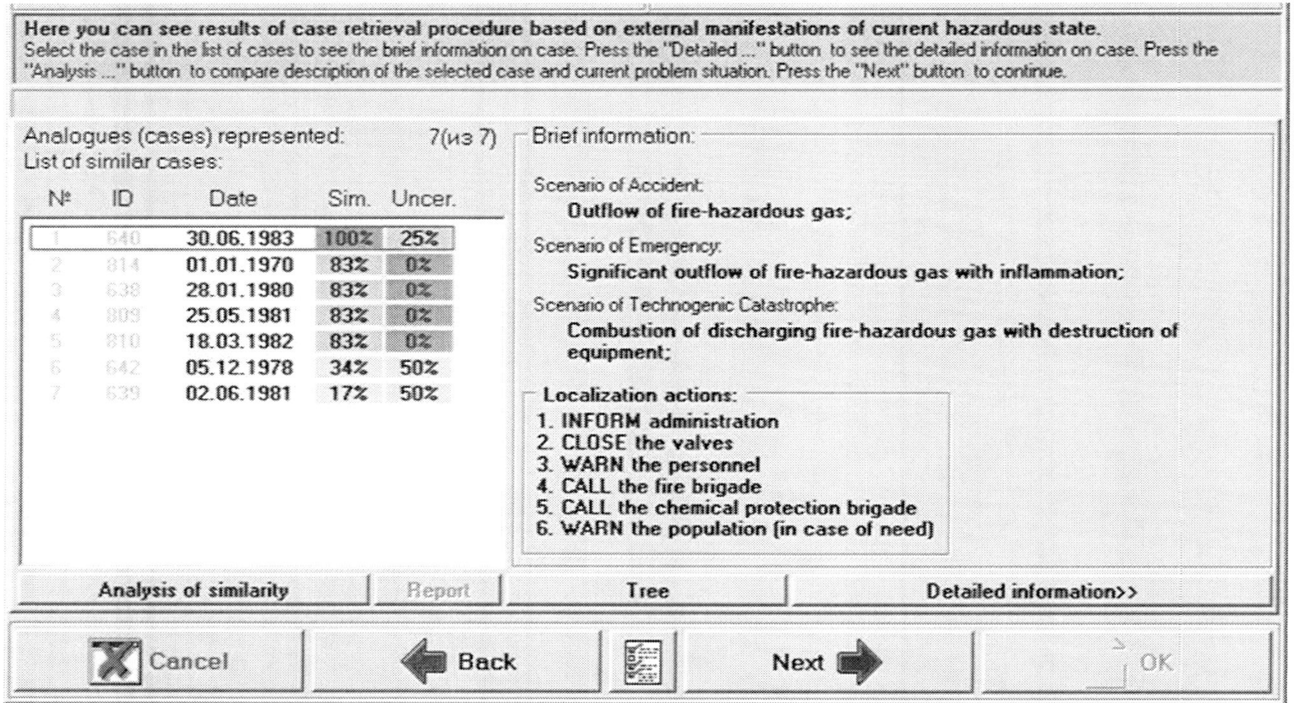

Figure 8. The example of graphic user interface of the software.

This TC has manifested itself as the event qualified as "Combustion of discharging fire-hazardous gas." Consider the application of the software to a given problem situation. The user inputs external manifestations of a current event and obtains (Figure 8):

- a list of possible scenarios related to the progress of the problem situation under scrutiny;
- the results related to forecasting of this problem situation are represented in the form of a event tree.
- the results of genesis of the situation are represented in the form of a failure tree.
- the list of scheduled operations and routine maintenances, which are needed for the purpose of localization and reducing any negative effects (consequences) of TC-stage.

CONCLUSION

We have completed the processes of conceptualization and formalization of data and knowledge, which reflect some changes in the state of hazardous technical systems. The hazardous states have been represented in the form of a cause-effect sequence of states: critical Failure (F), Accident (A), Emergency (E), Technogenic Catastrophe (TC).

The case-based model, which describes the proposed dynamics of hazardous states of a technological object, is developed. On its basis we have created a case base, which contains information about 200 failures and accidents that have taken place in the Soviet Union's (and later in Russia) chemistry and petrochemistry enterprises during the period from 1962 to 1996.

The algorithms and the case-based expert system (ES), which ensure the search decision process in a hierarchical space of cases and the process of adaptation of the decision on the basis of production rules and analytical models, are developed. In this ES, knowledge is represented (i) in the form of cases and (ii) in the form of production rules, which reflect the cause-effect complex related to the changes in the object's technical state.

For increasing the efficiency of the case retrieval procedure, we have applied the procedure of indexing of the cases. In the process of indexing, we have used elements of the object-oriented model, which includes a description of the hierarchical structure for the technological complex and a hierarchy of hazardous states.

A real example related to decision support in discovering the cause of an emergency with the aid of the software (intelligent system) has been considered.

The intelligent system elaborated has been used in investigations oriented to:

- revealing the factors, which condition the change in the technical state and safety of technological complex;
- grounding and planning the actions needed for minimizing the risk of occurrence of the undesirable critical states;
- rationalization of the actions oriented to restoring the states.

Noteworthy, the case-based approach cannot provide for finding a solution when there are no relevant cases in the case base. However, it may be used in combination with the knowledge represented in the form of production rules – as we have proposed above. Furthermore, a relevant case (or at least some very similar case) – when available – allows the expert to make some preliminary decision.

References

[1] Makhutov, N.A.; Petrov, V.P.; Achmetchanov, R.S.; Dubinin, E.F.; Dvoreckay, T.N. Features of scenario analysis for genesis and development of technogenic catastrophes. *Safety and Emergencies Problems.*2007, *3*, 3-28. (in Russia)

[2] Berman, A.F Formalization of formation processes of failure for unique mechanical systems. *Problems of Machinery Manufacture and Reliability.* 1994, *3*, 89-95.

[3] Berman, A.F.; Nikolaychuk, O.A. Structurization of investigation process for safety of complex technical systems. *Safety and Emergencies Problems.* 1999, *6*, 3-14. (in Russia)

[4] Berman, A.F.; Nikolaychuk, O.A. Modeling of investigation process for safety of complex technical systems. *Safety and Emergencies Problems.* 1999, *8*, 185-195. (in Russia)

[5] Berman, A.F.; Nikolaychuk, O.A.; Pavlov, A.I.; Yurin, A.Y. An intelligent system for decision support in the process of determination of causes of malfunctions and failures in the petro-chemical industry. *Automation in Industry.* 2006, 6, 15-17. (in Russia)

[6] Nikolaychuk, O.A.; Yurin, A.Y. Computer-aided identification of mechanical system's technical state with the aid of case-based reasoning. *Expert Syst Appl.* 2008, *34*, 635-642.

[7] Nikolaychuk, O.A. Automating studies of the technical state of dangerous mechanical systems. *Journal of Machinery Manufacture and Reliability.* 2008, *37(6)*, 597-602.

[8] Berman, A.F.; Nikolaychuk, O.A.; Pavlov, A.I.; Yurin, A.Y. An intelligent system for investigation and provision of safety for complex constructions. *International Journal Information Technologies & Knowledge.* 2008, *2 (3)*, 218-225.

[9] Jacobo, V.H.; Ortiz, A.; Cerrud, Y.; Schouwenaars, R. Hybrid expert system for the failure analysis of mechanical elements. *Eng Fail Anal.* 2007, *14(8)*, 1435-1443.

[10] Wang, H.C.; Wang, H.S. A hybrid expert system for equipment failure analysis. *Expert Syst Appl.* 2005, *28*, 615-622.

[11] Graham-Jones, P.; Mellor, B. Expert and knowledge-based systems in failure analysis. *Eng Fail Anal.* 1995, *2*, 137–149.

[12] Liao, T.W.; Zhang, Z.M.; Mount, C.R. A case-based reasoning system for identifying failure mechanisms. *Eng Appl Artif Intell.* 2000, *13*, 199–213.

[13] Liao, T.W. An investigation of a hybrid CBR method for failure mechanisms identification. *Eng Appl Artif Intell.* 2004, *17*, 123–134.

[14] Portinale, A.L.; Magro, D.; Torasso, P. Multi-modal diagnosis combining case-based and model-based reasoning: a formal and experimental analysis. *Artif Intell.* 2004, *158(2)*, 109-154.

[15] Jackson, P. *Introduction To Expert Systems*; Addison Wesley, 1998.

[16] Luger, G.F. *Artificial Intelligence: Structures and Strategies for Complex Problem*

Solving; Addison-Wesley, 2002.

[17] Aamodt, A.; Plaza, E. Case-based reasoning: foundational issues, methodological variations, and system approaches. *AI Communications*. 1994, *7(1)*, 39-59.

[18] Bergmann, R. Experience Management. *Lecture Notes on Artificial Intelligence*. 2002, *2432*.

[19] Booch, G. *Object-Oriented Analysis and Design with Applications*; Addison-Wesley, 1993.

[20] Berman, A.F.; Khramova, V.K. Automated data base for failures in pipelines and tubular high-pressure apparatus. *Chemical and Petroleum Engineering*. 1993, *29(2)*, 63-66.

[21] Berman, A.F.; Nikolaychuk, O.A. Technical state space of unique mechanical systems. *Journal of Machinery Manufacture and Reliability*. 2007, *36(1)*, 10-16.

[22] Zhuravlev, I. Yu.; Gurevitch, I. B. Pattern recognition and image recognition. In *Pattern recognition, classification, forecasting: Mathematical techniques and their application*; Zhuravlev, I. Yu. ; Ed.; Nauka: Moscow, 1989 ; Vol.2, pp 5-72. (in Russia)

[23] Ohsuga, S. *Knowledge processing*; Mir: Moscow, 1989. (in Russia)

[24] Plaza, E.; Arcos, J.L. Constructive adaptation. *Lecture Notes on Artificial Intelligence*. 2002, *2416*, 306-320.

[25] http://www.intersystems.com.

[26] http://www.borland.com/.

[27] www.ghg.net/clips/CLIPS.html

[28] Berman, A.F.; Moroz, V.G. Brittle fracture of pipes under the effect of external influences. *Strength Mater*. 1993, *25(2)*, 108-112.

In: Case-Based Reasoning: Processes Suitability…
Editors: Antonia M. Leeland

ISBN: 978-1-61728-352-9

Chapter 6

MATHEMATIZING THE CASE-BASED REASONING PROCESS

Michael Gr. Voskoglou *

Graduate Technological Educational Institute (T. E. I.), School of Technological Applications, Patras, Greece

ABSTRACT

In this chapter we introduce a finite absorbing Markov chain having as states the main steps of the Case-Based Reasoning (CBR) Process (retrieval, reuse, revision, and retaining), where retaining is the unique absorbing state. Applying standard results of the theory of finite Markov chains, we succeed in calculating the probabilities for the CBR process to be in a certain step at a certain phase of the solution of the corresponding real-world problem, and we obtain a measure for the effectiveness of the corresponding CBR system in solving similar new problems.

Next we present the first three of the above steps of the CBR process as fuzzy subsets in the set U of the linguistic labels of negligible, low, intermediate, high and complete success for each of the above steps. In this way, we build a fuzzy model for the representation of a CBR system, where we use the total possibilistic uncertainty as a measurement tool for its effectiveness in solving new related problems. Examples are also given to illustrate our results.

INTRODUCTION

As we have seen in the third section of the previous chapter, the main steps of the Case-Based Reasoning (CBR) process involve: R_1: *Retrieve* the most similar to the new problem past case, R_2: *Reuse* the information and knowledge of the retrieved case for the solution of the new problem, R_3: *Revise* the proposed solution, and R_4: *Retain* the part of this experience likely to be useful for future problem-solving.

---

* Corresponding author: E-mail: voskoglou@teipat.gr ; mvosk@tellas.gr.

According to the description of the CBR process given in the third section of the previous chapter one can design the "flow-diagram" of the CBR process shown in Figure 1.

In this chapter we shall attempt to give a mathematical formulation of the CBR process by using fundamental issues of the theory of finite Markov chains and of Fuzzy Sets.

The Markov Model

Roughly speaking a Markov chain is a stochastic process that moves in a sequence of phases through a set of states and has "no memory." This means that the probability of entering a certain state in a certain phase, although it is not necessarily independent of previous phases, depends *at most* on the state occupied in the previous phase. This property is known as the *Markov property*.

When its set of states is a finite set, then we speak about a *finite Markov chain*. For special facts on such type of chains we refer freely to Kemeny and Snell (1976).

We shall construct a Markov chain model for the mathematical description of the CBR process. For this, assuming that the CBR process has the Markov property, we introduce a finite Markov chain, having the four steps of the CBR process described in the previous section. The above assumption is a simplification (not far away from the truth) of the real system made in order to transfer from it to the "assumed real system." This is a standard technique applied during the mathematical modelling process of a real world problem, which enables the formulation of the problem in a form ready for mathematical treatment (Voskoglou, 2007, section 1).

Denoted by p_{ij} the transition probability from state R_i to R_j, for i,j=1,2,3,4, then the matrix A=[p_{ij}] is said to be the *transition matrix* of the chain.

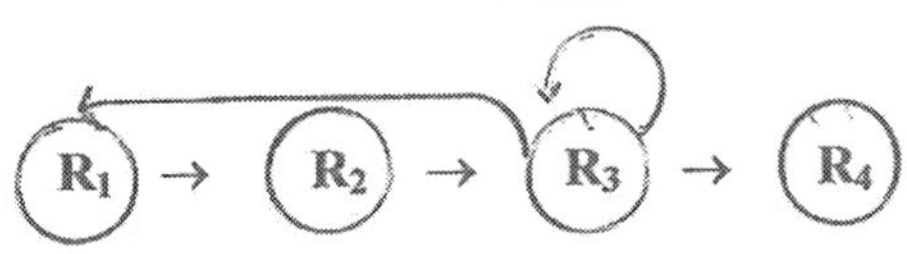

Figure 1. Flow-diagram of the CBR process

According to the flow-diagram of the CBR process shown in Figure 1, we find that

$$A = \begin{array}{c} \\ R_1 \\ R_2 \\ R_3 \\ R_4 \end{array} \begin{array}{c} \begin{array}{cccc} R_1 & R_2 & R_3 & R_4 \end{array} \\ \begin{bmatrix} 0 & 1 & 0 & 0 \\ 0 & 0 & 1 & 0 \\ p_{31} & 0 & p_{33} & p_{34} \\ 0 & 0 & 0 & 1 \end{bmatrix} \end{array},$$

where we obviously have that $p_{31}+p_{33}+p_{34}=1$.

Further let us denote by $\varphi_0,\varphi_1,\varphi_2,\ldots\ldots$ the successive phases of the above chain , and also denote by

$$P_i=[p_1^{(i)} p_2^{(i)} p_3^{(i)} p_4^{(i)}]$$

the row - matrix giving the probabilities $p_j^{(i)}$ for the chain to be in each of the states R_j, j=1,2,3,4, at phase φ_i, i=1,2,.... of the chain. We obviously have that

$$\sum_{j=1}^{4} p_j^{(i)} = 1.$$

The above row-matrix is called the *probability vector* of the chain at phase φ_i .

From the transition matrix A and the flow diagram of Figure 1 we obtain the "tree of correspondence" among the several phases of the chain and its states shown in Figure 2 . From this tree becomes evident that

$P_0 = [1\ 0\ 0\ 0]$, $P_1 = [0\ 1\ 0\ 0]$, $P_2 = [0\ 0\ 1\ 0]$, and $P_3 = [p_{31}\ 0\ p_{33}\ p_{34}]$.

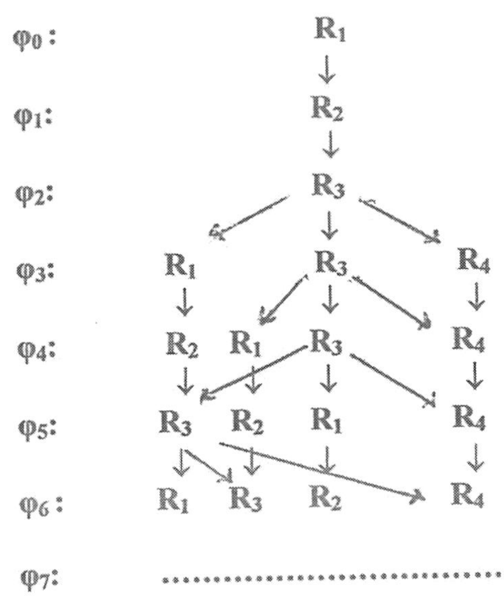

Figure 2. Tree of correspondence among states and phases of the Markov model

Further it is well known that

$$P_{i+1} = P_i A, \quad i=0,1,2,$$

Therefore we find that

$$P_4 = P_3 A = [p_{33}p_{31}\ \ p_{31}\ \ p_{33}^2\ \ p_{34}(p_{33}+1)]$$

$$P_5 = P_4 A = [p_{33}^2 p_{31}\ p_{33}p_{31}\ p_{31}+p_{33}^3\ p_{34}(p_{33}^2+p_{33}+1)]$$

and so on.

Observe now that, when the chain reaches state R_4, it is impossible to leave it, because the solution process of the new problem via the CBR approach finishes there. Thus, we have an *absorbing Markov chain* with R_4, its unique absorbing state. Applying standard techniques from the theory of finite absorbing chains, we bring the transition matrix A to its *canonical (or standard) form* A* by listing the absorbing state first and then partition it as follows:

$$A^* = \begin{array}{c} \\ R_4 \\ - \\ R_1 \\ R_2 \\ R_3 \end{array} \begin{array}{c} \begin{array}{cccc} R_4 & & R_1\ R_2\ R_3 \end{array} \\ \left[\begin{array}{c|ccc} 1 & 0 & 0 & 0 \\ - & - & - & - \\ 0 & 0 & 1 & 0 \\ 0 & 0 & 0 & 1 \\ p_{34} & p_{31} & 0 & p_{33} \end{array}\right] \end{array}$$

Symbolically we can write

$$A^* = \begin{bmatrix} I & | & 0 \\ - & | & - \\ R & | & Q \end{bmatrix},$$

where Q stands for the transition matrix of the non absorbing states. Then the *fundamental matrix* of the chain is given by

$$N = (I_3 - Q)^{-1} = \frac{adj(I_3 - Q)}{D(I_3 - Q)},$$

where I_3 denotes the 3X3 unitary matrix, adj(I_3-Q) denotes the adjoin matrix of I_3-Q and D(I_3-Q) denotes the determinant of I_3-Q. A straightforward calculation gives that

$$N = \frac{1}{1 - p_{31} - p_{33}} \begin{bmatrix} 1-p_{33} & 1-p_{33} & 1 \\ -p_{31} & 1-p_{33} & 1 \\ p_{31} & p_{31} & 1 \end{bmatrix} = [n_{ij}]$$

It is well known that the entry n_{ij} of N gives the mean number of times in state R_j when the chain is started in state R_i. Therefore, since the present chain is always starting from R_1, the sum

$$t = n_{11} + n_{12} + n_{13} = \frac{3 - 2p_{33}}{1 - p_{31} - p_{33}}$$

gives the mean number of phases of the chain before absorption. In other words, the mean number of steps for the completion of the CBR process is t+1. It becomes evident that, the bigger is the value of t, the greater is the difficulty encountered for the solution of the given problem via the CBR process. The ideal case is when the CBR process is completed straightforwardly, i.e. without "backwards" from R_3 to R_1, or "stays" to R_3 (see Figure 1). In this case we have that $p_{31}=p_{33}=0$ and $p_{34}=1$, therefore t=3. Thus, in general, we have that $t\geq 3$

.

The following simple example illustrates our results:

EXAMPLE: Consider the case of a physician, who takes into account the diagnosis and treatment of a previous patient having similar symptoms in order to determine the disease and treatment for the patient in front of him. Obviously the physician is using CBR. If the initial treatment fails to improve the health of the patient, then the physician either revises the treatment (this means stay to R_3 for two successive phases), or, in more difficult cases, is reminded of a previous similar failure and uses the failure case to improve his understanding of the present failure and correct it (this means transfer from R_3 to R_1). The process is completed when the physician succeeds in curing his patient.

Assume that the recorded statistical data show that the probabilities of a straightforward cure of the patient and of each of the above two reactions of the physician in case of failure are equal to each other. This means that $p_{13}=p_{33}=p_{34}=\frac{1}{3}$ and therefore t=7, i.e. the mean number of steps for the cure of the patient is 8.

Further, one finds that

$$P_3=[\frac{1}{3}\ 0\ \frac{1}{3}\ \frac{1}{3}],\ P_4=[\frac{1}{9}\ \frac{1}{3}\ \frac{1}{9}\ \frac{4}{9}],\ P_5=[\frac{1}{27}\ \frac{1}{9}\ \frac{4}{9}\ \frac{13}{27}]$$

and so on. Observing, for example, the probability vector P_5 one finds that the probability for the CBR process to be at the step of revision (R_3) in the 6th phase after its starting is $\frac{4}{9}$, or approximately *44.44%*, the corresponding probability to be at the step of retaining the acquired experience (R_4) is $\frac{13}{27}$, or approximately *48.15%* (in this case it is possible that the CBR process has arrived to the absorbing state R_4 in an earlier phase), etc.

Note: Knowing the exact "movements" during the CBR process, one can calculate the number of steps needed for the absorption of the chain directly from the flow-diagram of Figure 1. For example, considering the above case of the physician, assume that the initial treatment given to the patient failed to cure him and the physician was reminded of a similar failure in the past in order to correct it. Assume further that the new treatment didn't give the expected results and the physician revised it again, causing him to succeed in curing the patient. According to the assumptions mentioned above, it is easy, through the flow-diagram, to find that the number of steps needed for the absorption is exactly 8.

MEASURING THE EFFECTIVENESS OF A CBR SYSTEM

The challenge in CBR is to come up with methods that are suited for problem-solving and learning in particular subject domains and for particular application environments. In line with the main steps of the CBR process, core problems addressed by CBR research can be grouped into five areas: Representation of cases and methods for retrieval, reuse, revision and retaining the acquired experience. A CBR system should support the problems appearing in the above five areas. A good CBR system should support a variety of retrieval mechanisms and allow them to be mixed when necessary. In addition, the system should be able to handle large case libraries with the retrieval time increasing linearly (at worst) with the number of cases.

Let us consider a CBR system including a library of n recorded past cases and let t_i, as it has been calculated in the previous section, be the mean number of steps for the completion of the CBR process for case c_i, i=1,2,...,n. Each t_i could be stored in the system's library together with the corresponding case c_i. We define then the system's *effectiveness* (in solving new related problems), say t, to be the mean value of the t_i's of its stored cases, i.e. we have that

$$t = \frac{\sum_{i=1}^{n} t_i}{n}.$$

The more problems that are solved in future applications through the given system, the bigger becomes the number n of the stored cases in the system's library; therefore, the value of t is changing. As n increases it is normally expected that t will decrease, because the values of the t_i's of the new stored cases would be decreasing. In fact, the bigger is n, the better would be the chance of a new case to "fit" well (i.e. to have minor differences) with a known past case, and therefore the less would be the difficulty of solving the corresponding problem via the CBR process. Thus we could say that a CBR system "behaves well" if, when n tends to infinity, then its effectiveness t tends to 3.

EXAMPLE: Consider a CBR system that has been designed in terms of *Schank's model of dynamic memory* (Schank, 1982) for the representation of cases, as we have briefly described it in the second section of the previous chapter. In order to calculate the effectiveness of a system of this type we need first to calculate the effectiveness of each GE contained in it and then take the mean value of them.

For example, assume that the given system contains a GE including three cases, say c_1, c_2 and c_3. Assume further that c_1 corresponds to a straightforward successful application of the CBR process, that c_2 is the case described in the example of the third section of this chapter, and that c_3 includes one "return" from R_3 to R_1 and two "stays" to R_3. Then t_1=3 and t_2=7, while for the calculation of t_3 observe that $p_{31}=p_{34}=\frac{1}{4}$ and $p_{33}=\frac{1}{2}$, therefore t_3=8. Thus the effectiveness of this GE is equal to $t = \frac{3+7+8}{3} = 6$. Notice that a complex GE may contain some more specific GE's (e.g. see Figure 3 in page 12 of Aamodt and Plaza, 2004). In this case we only need to calculate the efficiency of the complex GE by considering all its cases, regardless if they belong or not to one or more of the specific GE's contained in it.

Note: As we have seen in the second section of the previous chapter, an alternative approach for the representation of cases in a CBR system is the *category and exemplar model* (Porter and Bareiss, 1986).The process of calculating the effectiveness of a system of such type is analogous to the process described in the above example, the only difference being that one has to work with categories instead of GE's. In a similar way one may calculate the effectiveness of systems corresponding to other case memory models that we have seen in the second section of the [previous?] chapter, including Rissland's (1983) and Ashley's HYPO, the MBR model of Stanfill and Waltz (1988), etc.

FUZZY SETS

They are often situations in everyday life in which definitions have not clear boundaries,; e.g. this happens when we speak about the "high mountains" of a country, the "good players" of a football team, etc . The fuzzy sets theory was created in response to have a mathematical representation of such kind of situations.

Let U denote the universal set. Then a *fuzzy subset* A of U (called often for simplicity a fuzzy set in U), initiated by Zadeh (1965), is defined in terms of the membership function m_A that assigns to each element of U a real value from the interval [0,1]. In more specific terms

$$A = \{(x, m_A(x)) : x \in U\}$$

where $m_A : U \rightarrow [0,1]$.

The value $m_A(x)$, called the *membership degree (or grade) of x in A*, expresses the degree to which x verifies the characteristic property of A. The nearer is the value $m_A(x)$ to 1, the higher is the membership degree of x in A.

Despite the fact that they can take on similar values, it is important to realize that membership degrees are not probabilities. One immediately apparent difference is that the summation of probabilities on a universal set must equal 1, while there is no such requirement for membership degrees.

The methods of choosing the suitable membership function for each case are usually empiric, based on experiments made on a sample of the population that we study.

Obviously each classical (crisp) subset A of U may be considered as a fuzzy subset of U, with

$$m_A(x)=1$$

if $x \in U$ and

$$m_A(x)=0$$

if $x \notin U$.

Most of the concepts of classical (crisp) sets can be extended in terms of the above definition to fuzzy sets. For example, if A and B are fuzzy sets in U, then A is called a *subset of B* if

$$m_A(x) \leq m_B(x)$$

for each x in U, while the *intersection* $A \cap B$ is a fuzzy subset of U with membership function

$$m_{A \cap B}(x) = \min \{ m_A(x), m_B(x) \}$$

Further, given a positive integer, say n, a fuzzy subset of the cartesian product U^n is called a *fuzzy relation in U*, etc.

For special facts on fuzzy sets and on uncertainty theory we refer freely to Klir and Folger (1988).

A Fuzzy Model for the Representation of a CBR System

Let us consider a CBR system whose library contains n past cases, $n \geq 2$. We denote by R_i , i=1,2,3 , the steps of retrieval, reuse and revision and by a, b, c, d, and e the linguistic labels of negligible, low, intermediate, high and complete degree of success respectively for each of the R_i's. Set

$$U=\{a, b, c, d, e\}$$

We are going to represent R_i's as fuzzy sets in U. For this, if n_{ia}, n_{ib}, n_{ic}, n_{id} and n_{ie} respectively denote the number of cases where it has been achieved negligible, low, intermediate, high and complete degree of success for the state R_i i=1,2,3, we define the membership function m_{Ri} in terms of the frequencies, i.e. by

$$m_{Ri}(x)=\frac{n_{ix}}{n}$$

for each x in U. Thus we can write

$$R_i= \{(x, \frac{n_{ix}}{n}) : x \in U\}, i=1,2,3$$

The reason, for which we didn't include the last step R_4 of the CBR process in our fuzzy representation, is that all past cases, either successful, or not, are retained in the system's library and therefore there is no fuzziness in this case. In other words keeping the same notation we have that $n_{4a}=n_{4b}=n_{4c}=n_{4d}=0$ and $n_{4e}=1$.

In order to represent all possible *profiles (overall states)* of a case during the CBR process, we consider a fuzzy relation, say R, in U^3 of the form

$$R=\{(s, m_R(s)) : s=(x, y, z) \in U^3\}$$

To determine properly the membership function m_R we give the following definition:

DEFINITION: A profile s=(x, y, z), with x, y ,z in U, is said to be ***well ordered*** if x corresponds to a degree of success equal or greater than y, and y corresponds to a degree of success equal or greater than z.

For example, profile (c, c, a) is well ordered, while (b, a, c) is not.

We define now the membership degree of s to be

$$m_R(s)=m_{R_1}(x)m_{R_2}(y)m_{R_3}(z)$$

if s is a well ordered profile, and zero otherwise. In fact, if for example (b, a, c) possessed a nonzero membership degree, given that the degree of success at the step of reuse is negligible how the proposed solution could be revised?

In order to simplify our notation we shall write m_s instead of $m_R(s)$. Then the *possibility* r_s of the profile s is given by

$$r_s=\frac{m_s}{\max\{m_s\}}$$

where $\max\{m_s\}$ denotes the maximal value of m_s , for all s in U^3. In other words r_s is the "relative membership degree" of s with respect to the other profiles.

During the CBR process it might be used reasoning that involves amplified inferences, whose content is beyond the available evidence and hence obtain conclusions not entailed in the given premises. The appearance of conflict in the conclusions requires that the conclusions be appropriately adjusted so that the resulting generalization is free of conflict. The value of total conflict during the CBR process can be measured by the *strife function* S(r) on the ordered possibility distribution

$$r:\ r_1=1\geq r_2 \geq \ldots\ldots \geq r_n \geq r_{n+1}$$

of the profiles defined by:

$$S(r) = \frac{1}{\log 2}\left[\sum_{i=1}^{n}(r_i - r_{i+1})\log\frac{i}{\sum_{j=1}^{i} r_j}\right]$$

In general, the amount of information obtained by an action can be measured by the reduction of uncertainty that results from the action. Thus the *total possibilistic uncertainty* T(r) during the CBR process can be used as a measure for the system's effectiveness in solving new related problems. The value of T(r) is measured by the sum of the strife S(r) and *non specificity* N(r) (Klir, 1995; p.28), defined by:

$$N(r) = \frac{1}{\log 2}[\sum_{i=2}^{n}(r_i - r_{i+1})\log i$$

In contrast to strife, which, as we have already seen, expresses conflicts among the various sets of alternatives, non specificity is connected with the sizes (cardinalities) of relevant sets of alternatives. The lower is the value of T(r), the higher is the effectiveness of the CBR system in solving new related problems.

Assume now that one wants to study the combined results of the behavior of k different systems, $k \geq 2$, designed for the solution of the same type of problems via the CBR process. Then it becomes necessary to introduce the *fuzzy variables* $R_1(t)$, $R_2(t)$ and $R_3(t)$ with t=1,2,…,k. The values of the above variables represent the steps of the CBR process for each of the k CBR systems as fuzzy sets in U; e. g. $R_1(2)$ represents the step of retrieval for the second system.. In order to measure the degree of evidence of the combined results of the k systems, it becomes necessary to define the possibility r(s) of each profile s with respect to the sum of the membership degrees of s for all systems. For this, we introduce the *pseudo-frequencies*

$$f(s) = \sum_{t=1}^{k} m_s(t)$$

and we define

$$r(s) = \frac{f(s)}{\max\{f(s)\}}$$

where max{f(s)} denotes the maximal pseudo-frequency. Obviously the same method could be applied when one wants to study the behaviour of a system during the CBR process for the solution of k different related problems.

An Application of the Fuzzy Model

Let us consider a CBR system with an existing library of 105 past cases, where in no case there was a failure at the step of retrieval of a past case for the solution of the corresponding problem. More explicitly, let us assume that in 51 cases we had an intermediate success in retrieving a suitable past case, in 24 cases high, and in 30 cases we had a complete success respectively. Of course the existence of a certain criterion is necessary in order to be able to characterize the degree of success of retrieval for each of the past cases. Thus the step of retrieval can be represented as a fuzzy set in U as

$$R_1 = \{(a,0),(b,0),(c,\tfrac{51}{105}),(d,\tfrac{24}{105}),(e,\tfrac{30}{105})\}$$

Assume further that in a similar way we obtained that

$$R_2 = \{(a,\tfrac{18}{105}),(b,\tfrac{18}{105}),(c,\tfrac{48}{105}),(d,\tfrac{21}{105}),(e,0)\}$$

and

$$R_3 = \{(a,\tfrac{36}{105}),(b,\tfrac{30}{105}),(c,\tfrac{39}{105}),(d,0),(e,0)\}$$

It is a straightforward process now to calculate the membership degrees of all the possible profiles (see column of $m_s(1)$ in Table 1). For example, if

$$s=(c, b, a)$$

then

$$m_s = m_{R_1}(c).m_{R_2}(b).m_{R_3}(a) = \frac{51}{105}\frac{18}{105}\frac{36}{105} \approx 0{,}029 .$$

It turns out that (c, c, c) is the profile with the maximal membership degree 0,082 and therefore the possibility of each s in U^3 is given by

$$r_s = \frac{m_s}{0{,}082}$$

For example the possibility of (c, b, a) is

$$\frac{0{,}029}{0{,}082} \approx 0{,}353$$

while the possibility of (c, c, c) is of course equal to 1.

Calculating the possibilities of the

$$5^3=125$$

in total profiles (see column of $r_s(1)$ in Table 1) one finds that the ordered possibility distribution r of the profiles is:

$$r_1=1,\ r_2=0{,}92,\ r_3=0{,}768,\ r_4=0{,}512,\ r_5=0{,}476,$$

$$r_6=0{,}415,\ r_7=0{,}402,\ r_8=0{,}378,\ r_9=r_{10}=0{,}341,\ r_{11}=0{,}329,$$

r_{12}=0,317, r_{13}=0,305, r_{14}=0,293, $r_{15}=r_{16}$=0,256, r_{17}=0,20, r_{18}=0,195,

r_{19}=0,171, $r_{20}=r_{21}=r_{22}$=0,159, r_{23}=0,134, $r_{24}=r_{25}=$ =........$=r_{125}$=0

Therefore the total possibilistic uncertainty of the system is

$$T(r)=S(r)+N(r)=0,565+2,405= =2,97$$

Next we shall study the combined results of the behavior of the above system and of another system, designed for the solution of the same type of problems via the CBR process, with an existing library of 90 past cases. Assume that working as before we found for the second system that

$$R_1=\{(a,0),(b,\tfrac{18}{90}),(c,\tfrac{45}{90}),(d,\tfrac{27}{90}),(e,0)\}$$

$$R_2=\{(a,\tfrac{18}{90}),(b,\tfrac{24}{90}),(c,\tfrac{48}{90}),(d,0),(e,0)\}$$

and

$$R_3=\{(a,\tfrac{36}{90}),(b,\tfrac{27}{90}),(c,\tfrac{27}{90}),(d,0),(e,0)\}$$

The calculation of all possible profiles gives the results shown in column of $m_s(2)$ in Table 1. It turns out that (c, c, a) is the profile possessing the maximal membership degree 0,107 and therefore the possibility of each s is given by

$$r_s=\frac{m_s}{0,107}$$

(see column of $r_s(2)$ in Table 1).

Finally, in the same way as above, one finds that

$$T(r)=S(r)+N(r)=0,452+1,87= =2,322$$

Thus, since

$$2,322<2,97$$

the effectiveness of the second system in solving new related problems is better than that of the first one. This happens despite to the fact that the profile (c, c, c) with the maximal possibility of appearance in the first system is a more satisfactory profile than the corresponding profile (c, c, a) of the second system.

Notice that in general, the more are the stored past cases in the system's library, the greater is expected to be its effectiveness in solving new related problems. In fact, the more are the past cases, the greater is the probability for a new problem to fit satisfactorily to one of them. Therefore the fact that the second system was found to be more effective than the first one, although not impossible to happen, it is rather unexpected in general.

Table 1. Student profiles with non zero pseudo-frequencies (the outcomes are of accuracy up to the third decimal point)

A_1	A_2	A_3	$m_s(1)$	$r_s(1)$	$m_s(2)$	$r_s(2)$	f(s)	r(s)
b	b	b	0	0	0,016	0,150	0,016	0,087
b	b	a	0	0	0,021	0,196	0,021	0,115
b	a	a	0	0	0,016	0,150	0,016	0,087
c	c	c	0,082	1	0,080	0,748	0,162	0,885
c	**c**	**a**	**0,076**	**0,927**	**0,107**	**1**	**0,183**	**1**
c	c	b	0,063	0,768	0,008	0,075	0,071	0,388
c	a	a	0,028	0,341	0,040	0,374	0,068	0,372
c	b	a	0,028	0,341	0,053	0,495	0,081	0,443
c	b	b	0,024	0,293	0,040	0,374	0,064	0,350
d	d	a	0,016	0,495	0	0	0,016	0,087
d	d	b	0,013	0,159	0	0	0,013	0,074
d	d	c	0,021	0,256	0	0	0,021	0,115
d	a	a	0,013	0,159	0,024	0,224	0,037	0,202
d	b	a	0,013	0,159	0,032	0,299	0,045	0,246
d	b	b	0,011	0,134	0,024	0,224	0,035	0,191
d	c	a	0,031	0,378	0,064	0,598	0,095	0,519
d	c	b	0,026	0,317	0,048	0,449	0,074	0,404
d	c	c	0,034	0,415	0,048	0,449	0,082	0,448
e	a	a	0,017	0,207	0	0	0,017	0,093
e	b	b	0,014	0,171	0	0	0,014	0,077
e	c	a	0,039	0,476	0	0	0,039	0,213
e	c	b	0,033	0,402	0	0	0,033	0,180
e	c	c	0,042	0,512	0	0	0,042	0,230
e	d	a	0,025	0,305	0	0	0,025	0,137
e	d	b	0,021	0,256	0	0	0,021	0,115
e	d	c	0,027	0,329	0	0	0,027	0,148

We introduce now the fuzzy variables $R_1(t)$, $R_2(t)$, and $R_3(t)$, with t=1,2. Then the pseudo-frequency of each profile s is given by

$$f(s)=m_s(1)+m_s(2)$$

(see the corresponding column of Table 1). It turns out that (c, c, a) is the profile with the highest pseudo-frequency 0,183 and therefore the possibility of each profile is given by

$$r(s)=\frac{f(s)}{0{,}183}$$

The possibilities of all profiles having nonzero pseudo-frequencies are given in the last column of Table 1

CONCLUSION

The following can be drawn from the discussion presented in this chapter:

- Our *Markov model* was built by introducing a finite, absorbing Markov chain, having as stated, the major steps of the CBR process, i.e. the actions retrieve, reuse, revise and retain. The starting step is always the first of the above actions, while the unique absorbing state is the last one. Thus, by applying standard results of the theory of finite Markov chains, we succeeded in calculating the probabilities of the CBR process to be in a certain step at any given phase of the solution of the corresponding problem, and we also obtained a measure for the effectiveness of the corresponding CBR system.
- Our *fuzzy model* for the description of the CBR system in use was obtained by representing the steps of retrieval, reuse and revision of the CBR process as fuzzy subsets of the set U of the linguistic labels of negligible, low, intermediate, high and complete success of each of the above steps. In this fuzzy environment, we used the total possibilistic uncertainty as a measurement tool for the effectiveness of the CBR system in solving new related problems.
- The fuzzy model is not restricted in giving quantitative information only (possibilities, value of T(r), etc), but it also gives a qualitative view of the behavior of the CBR system. In fact, through this, one studies all the possible profiles of the stored cases, and gets – in terms of the linguistic labels – a comprehensive idea about the degree of success in each step of the CBR process.
- Another advantage of the fuzzy model is that it gives the possibility for studying the combined results of the behavior of two or more CBR systems designed for the solution of the same type of problems, or, alternatively, the behaviour of the CBR system in use during the solution of two, or more, new related problems.
- In an earlier paper (Voskoglou, 2008), an analogous model was constructed for the fuzzy representation of the process of learning a subject matter by a group of students in the classroom. Analogous efforts to use the fuzzy logic in the area of student modelling, and student diagnosis in particular, and in education in general, have been attempted by other researchers as well, e.g. Perikaris (1996), Espin and Oliveras (1997), Ma and Zhou (2000), Ajello and Spagnolo (2002), Spagnolo and Gras (2004), etc.
- Markov chain models were also used in earlier papers of the author as an alternative approach for the same or analogous purposes, e.g. Voskoglou and Perdikaris (1991), Voskoglou (1996a, 1996b, 2000, 2007) etc. However, Markov models, although easier sometimes to be applied in practice by a non expert, are self- restricted to provide quantitative information only for the corresponding situations, e.g. measures for the problem-solving, or model-building abilities of a group of students, short and long-run forecasts (probabilities) for the evolution of various phenomena, etc. Therefore, one could claim that a fuzzy model, like the one presented in this paper, is more useful for the deeper study of a real situation, because, apart from the quantitative information, it gives also the possibility of a qualitative analysis of the corresponding phenomena.

- Although we have presented some examples in both cases (Markov and fuzzy model) to illustrate our results, further study and research are needed in order to estimate the usefulness of these results in practical applications of real world problems and of situations of our everyday life.

REFERENCES

Aamodt, A. & Plaza, E. (1994). Case-Based Reasoning:: Foundational Issues, Methodological Variations, and System Approaches, *Artificial Intelligence Communications*, *7/1*, 39-52.

Ajello, M. & Spagnolo, F. (2002). *Some experimental observations on common sense and fuzzy logic*, Proceedings International Conference on Mathematics Education into the 21st Century, 35-39, Palermo, Italy.

Espin, E. A. & Oliveras, C. M. L. (1997). Introduction to the use of fuzzy logic in the assessment of mathematics teachers'. In A. Gagatsis, (Ed), *Proceedings of the 1st Mediterranean Conference on Mathematics Education (MEDCONF 97)*, 107-113, Nicosia, Cyprus.

Kemeny, J. & Snell, J. l. (1976). *Finite Markov Chains*, Springer-Verlag, New York.

Klir, G. J. & Folger, T. A. (1988). Fuzzy Sets, *Uncertainty and Information*, Prentice Hall, London.

Klir, G. J. (1995). Principles of uncertainty: What are they? Why do we need them?", *Fuzzy Sets and Systems*, *74*, 15-31.

Ma, J. & Zhou, D. (2000). Fuzzy Set Approach to the Assessment of Student-Centered Learning, *IEEE Transactions on Education*, *43(2)*, 237-241.

Perdikaris, S. (1996). A system framework for fuzzy sets in the van Hiele level theory of geometric reasoning, *International Journal of Mathematical Education in Science and Technology*, *27(2)*, 273-278.

Porter, B. & Bareiss, B. (1986). PROTOS: An experiment in knowledge acquisition for heuristic classification tasks. *In Proceedings of the First International Meeting on Advances in Learning (IMAL)*, 159-174, Les Arcs, France.

Rissland, E. (1983). Examples in legal reasoning: Legal hypotheticals. *In: Proceedings 8th International Joint Conference on Artificial Intelligence (IJCAI)*, Karlsruhe.

Schank, R. (1982). Dynamic memory: *A theory of reminding and learning in computers and people*, Cambridge Univ. Press.

Spagnolo, F. & Gras, R. (2004). Fuzzy implication through statistic implication: A new approach in Zadeh's framework. In D., Scott, L., Kurgan, P., Musilek, W. Pedrycz, & M. Reformat, (Eds), 23d International Conference of the North American Fuzzy Information Processing Society (NAFIPS 2004), *IEEE*, Vol.*1*, 425-429.

Stanfill, C. & Waltz, D. (1988). The memory-based reasoning paradigm. In: *Case-based reasoning*, Proceedings from a workshop, 414-424, Morgan Kaufmann, Clearwater Beach, Florida.

Voskoglou, M. Gr. & Perdikaris, S. (1991). A Markov chain model in problem-solving, *International Journal of Mathematical Education in Science and Technology*, *24(3)*, 443-447.

Voskoglou, M. Gr. (1996 a). The use of Markov chains to describe the process of learning, Theta: *A Journal of Mathematics (Manchester Metropolitan University*, UK), *10(1)*, 36-40.

Voskoglou, M. Gr. (1996 b). An application of ergodic Markov chains to analogical problem solving, *The Mathematics Education (India)*, Vol. *XXX (2)*, 95-108.

Voskoglou, M. Gr. (2000). *An application of Markov chains to decision making,* Studia Kupieckie, (University of Lodz, Poland), *6*, 69-76.

Voskoglou, M. Gr. (2007). A stochastic model for the modelling process. In Chr., Haines, P., Galbraith, W. Bloom, & S. Khan, (Eds), *Mathematical Modelling: Education, Engineering and Economics (ICTMA12)*, 149-157, Horwood, Chichester, UK.

Voskoglou, M. Gr. (2009). Fuzziness or probability in the process of learning? A general question illustrated by examples from teaching mathematics, *The Journal of Fuzzy Mathematics*, *17(3)*, 79-686, International Fuzzy Mathematics Institute (Los Angeles).

Zadeh, L. A. (1965). Fuzzy Sets, *Information and Control*, *8*, 338-353.

In: Case-Based Reasoning: Processes Suitability…
Editors: Antonia M. Leeland
ISBN: 978-1-61728-352-9

Chapter 7

PROTOTYPE-BASED REASONING FOR DIAGNOSIS OF DYSMORPHIC SYNDROMES

Rainer Schmidt*

Institute for Medical Informatics and Biometry,
University of Rostock, Germany

ABSTRACT

In Case-Based Reasoning, usually former single cases are considered. For medical diagnosis, we propose a method that does not retrieve single but generalised, prototypical cases (prototypes).

Since diagnosis of dysmorphic syndromes is a domain with incomplete knowledge and where even experts have seen only few syndromes themselves during their lifetime, documentation of cases and the use of case-oriented techniques are popular. So, most of the systems dealing with the diagnosis of dysmorphic syndromes, perform classification based on prototypes. Different prototypicality measures are applied to determine the most probable syndrome. These measures differ from the usual Case-Based Reasoning similarity measures, because here cases and syndromes are not represented as attribute value pairs but as long lists of symptoms, and because query cases are not compared with cases but with prototypes.

In contrast to other dysmorphic systems our approach additionally applies adaptation rules. These rules do not only consider single symptoms but combinations of them, which indicate high or low probabilities of specific syndromes. However, it is a big problem to acquire such domain dependent adaptation rules. First, some medical experts, with difficulties, provided a few rules. Recently, we have generated suggestions for adaptation rules automatically and have discussed them subsequently with medical experts.

* Corresponding author: rainer.schmidt@uni-rostock.de.

1. INTRODUCTION

Knowledge of physicians consists of general knowledge they have read in medical books and of their experiences in form of cases they have treated themselves or colleagues have told them about. Some cases are typical whereas others are rather exceptional. Doctors consider differences between their current patient and typical or known exceptional cases. We believe that medical Case-Based Reasoning (CBR) systems should take the reasoning of doctors into account [1]. Such systems should not only consist of general medical domain knowledge plus a flat case base, but the case base should be structured by typical case generalisations called prototypes [2].

Though the use of prototypes had been early introduced in the CBR community [3, 4], their use is still rather seldom. Later on it fell into oblivion and was brought up again by Bergman in form of generalised cases [5]. His notion of generalised cases is similar but not identical to our idea of prototypes. Whereas generalised cases are general or abstract in contrast to concrete cases, prototypes contain the typical features of a set of cases.

However, since doctors reason with typical cases anyway, in medical CBR systems prototypes are a rather common knowledge form, they are used in a variety of applications, e.g. for diabetes [6], for eating disorders [7], and for pulmonology [8].

A Prototype is generalised from a set of single cases. The cases in this set are very similar to each other or they belong in some other specific way together and form a sort of class. For example, in a diagnostic system all patients diagnosed as measles patient might be grouped together. Usually, prototypes have the same structure as cases but have less and more general features, namely just the typical ones. Sometimes prototypes are defined by medical experts, sometimes they can be found in literature (e.g. the typical symptoms for measles), and sometimes they are computed. The use of case oriented generalised knowledge presents the opportunity to structure case bases. Cases can be clustered into groups, prototypical diseases or schema. Clancey [9] distinguishes between prototypes that represent specific expressions of diseases or therapies and schema that contain essential features of diseases or therapies. As Selz [10] characterises a schema as a description of an entity where at least one part remains vague, the distinction between prototypes and schema seems to be fluid. We only use the term prototype and refer to a hierarchy of prototypes where the most general prototypes that contain the most common features are situated on top and the most specific ones are placed at the bottom. This notion of prototypes differs from the usual notion of classes and clusters [11] in many ways. Prototypes are not the result of a classification process. Whether a case belongs to a prototype, is determined by its features or defined by an expert. There may be a hierarchy of prototypes but there are not relations (similarity, is-a and so on), and the set of cases belonging to a prototype is not represented by its most representative case but by the prototype. The main purpose of such generalised knowledge is to guide the retrieval and sometimes to decrease the amount of storage by erasing redundant cases. In domains with rather weak domain theories another advantage of case-oriented techniques is their ability to learn from cases. Only gathering new cases may improve the systems ability to find suitable similar cases for current problems, but it does not elicit the intrinsic knowledge of the stored cases. To learn the knowledge contained in cases a generalisation process is necessary. Generally speaking, prototypes fill the knowledge gap between the specificity of single cases and abstract knowledge usually expressed as rules.

In this chapter, we present a system that even goes a step further. Prototypes are not just to structure the case base but they are even used for reasoning instead of single cases.

1.1. Diagnostic Support for Dysmorphic Syndromes

When a child is born with dysmorphic features or with multiple congenital malformations or if mental retardation is observed at a later stage, finding the correct diagnosis is extremely important. Knowledge of the nature and the etiology of the disease enables the pediatrician to predict the patient's future course. So, an initial goal for medical specialists is to diagnose a patient to a recognised syndrome. Genetic counselling and a course of treatments may then be established.

A dysmorphic syndrome describes a morphological disorder and it is characterised by a combination of various symptoms, which form a pattern of morphologic defects. An example is Down Syndrome which can be described in terms of characteristic clinical and radiographic manifestations such as mental retardation, sloping forehead, a flat nose, short broad hands and generally dwarfed physique [12].

The main problems of diagnosing dysmorphic syndromes are as follows [13]:

- more than 200 syndromes are known,
- many cases remain undiagnosed with respect to known syndromes,
- usually many symptoms are used to describe a case (between 40 and 130),
- every dysmorphic syndrome is characterised by nearly as many symptoms.

Furthermore, knowledge about dysmorphic disorders is continuously modified, new cases are observed that cannot be diagnosed (it exists even a journal that only publishes reports of observed interesting cases [14]), and sometimes even new syndromes are discovered. Usually, even experts of paediatric genetics only see a small count of dysmorphic syndromes during their lifetime.

So, we have developed a diagnostic system that uses a large case base. Starting point to build the case base was a large case collection of the paediatric genetics of the University of Munich, which consists of nearly 2000 cases and 229 prototypes. A prototype (prototypical case) represents a dysmorphic syndrome by its typical symptoms. Most of the dysmorphic syndromes are already known and have been defined in the literature. And nearly one third of our entire case base has been determined by semiautomatic knowledge acquisition, where an expert selected cases that should belong to same syndrome and subsequently a prototype, characterised by the most frequent symptoms of his cases, was generated. To this database we have added cases from "clinical dysmorphology" [14] and syndromes from the London dysmorphic database [15], which contains only rare dysmorphic syndromes.

1.2. Other Systems

Systems to support diagnosis of dysmorphic syndromes have already been developed in the early 80's. The simple ones perform just information retrieval for rare syndromes, namely the London dysmorphic database [14], where syndromes are described by symptoms, and the Australian POSSUM, where syndromes are visualised [16]. Diagnosis by classification is done in a system developed by Wiener and Anneren [17]. They use more than 200 syndromes as database and apply Bayesian probability to determine the most probable syndromes. Another diagnostic system, which uses data from the London dysmorphic database was developed by Evans [18]. Though he claims to apply Case-Based Reasoning, in fact it is again just a classification, this time performed by Tversky's measure of dissimilarity [19]. The most interesting aspect of his approach is the use of weights for the symptoms. That means the symptoms are categorised in three groups – independently from the specific syndromes, instead only according to their intensity of expressing retardation or malformation. However, Evans admits that even features, that are usually unimportant or occur in very many syndromes sometimes play a vital role for discrimination between specific syndromes.

The novelty of our approach is that we do not only perform classification but subsequently apply adaptation rules. These rules do not only consider single symptoms but specific combinations of them, which indicate high or low probabilities of specific syndromes.

1.3. Case-Based Reasoning and Prototypicality Measures

Since the idea of Case-Based Reasoning (CBR) is to use former, already solved solutions (represented in form of cases) for current problems [20], CBR seems to be appropriate for diagnosis of dysmorphic syndromes. CBR consists of two main tasks [21], namely retrieval, which means searching for similar cases, and adaptation, which means adapting solutions of similar cases to the query case. For retrieval usually explicit similarity measure or, especially for large case bases, faster retrieval algorithms like Nearest Neighbour Matching [22] are applied. For adaptation only few general techniques exist [23], usually domain specific adaptation rules have to be acquired.

In CBR usually cases are represented as attribute-value pairs. In medicine, especially in diagnostic applications, this is not always the case, instead often a list of symptoms describes a patient's disease. Sometimes these lists can be very long, and often their lengths are not fixed but vary with the patient. For dysmorphic syndromes usually between 40 and 130 symptoms are used to characterise a patient.

Furthermore, for dysmorphic syndromes it is unreasonable to search for single similar patients (and of course none of the systems mentioned above does so) but for more general prototypes that contain the typical features of a syndrome. Prototypes are a generalisation from single cases. They fill the knowledge gap between the specificity of single cases and abstract knowledge in the form of cases.

So, to determine the most similar prototype for a given query patient instead of a similarity measure a prototypicality measure is required. One speciality is that for prototypes the list of symptoms is usually much shorter than for single cases.

The result should not be just the one and only most similar prototype, but a list of them – sorted according to their similarity. So, the usual CBR methods like indexing or nearest neighbour search are inappropriate. Instead, rather old measures for dissimilarities between concepts [8, 24] are applied and explained in the next section.

2. DIAGNOSIS OF DYSMORPHIC SYNDROMES

Our system consists of four steps. At first the user has to select the symptoms that characterise a new patient. This selection is a long and very time consuming process, because we consider more than 800 symptoms. However, diagnosis of dysmorphic syndromes is not a task where the result is very urgent, but it usually requires thorough reasoning and afterwards a long-term therapy has to be started. Optionally, the user may select a prototypicality measure. If he/she does select a measure, the measure proposed by Tversky is used. At present there are three choices. As humans look upon cases as more typical for a query case as more features they have in common [24], distances between prototypes and cases usually mainly consider the shared features.

2.1. Prototypicality Measures

The first, rather simple measure (1) just counts the number of matching symptoms of the query patient (X) and a prototype (Y) and normalises the result by dividing it by the number of symptoms characterising the syndrome.

This normalisation is done, because the lengths of the lists of symptoms of the various prototypes vary very much. It is performed by the two other measures too.

The following equations are general (as they were originally proposed) at the point that a general function "f" is used, which usually means a sum that can be weighted. In general these functions "f" can be weighted differently. However, since we do not use any weights at all, in our application "f" means simply a sum.

$$D(X,Y) = \frac{f(X+Y)}{f(Y)} \qquad (1)$$

The second measure (2) was developed by Tversky [19]. It is a measure of dissimilarity for concepts. In contrast to the first measure, additionally two numbers are subtracted from the number of matching symptoms. Firstly, the number of symptoms that are observed for the patient but are not used to characterise the prototype (X-Y), and secondly the number of symptoms used for the prototype but are not observed for the patient (Y-X) is subtracted.

$$D(X,Y) = \frac{f(X+Y) - f(X-Y) - f(Y-X)}{f(Y)} \qquad (2)$$

The third prototypicality measure (3) was proposed by Rosch and Mervis [24]. It differs from Tversky's measure only in one point: the factor X-Y is not considered:

$$D(X,Y) = \frac{f(X+Y) - f(Y-X)}{f(Y)} \qquad (3)$$

In the third step to diagnose dysmorphoic syndromes, the chosen measure is sequentially applied on all prototypes (syndromes). Since the syndrome with maximal similarity is not always the right diagnosis, the 20 syndromes with best similarities are listed in a menu (figure 1).

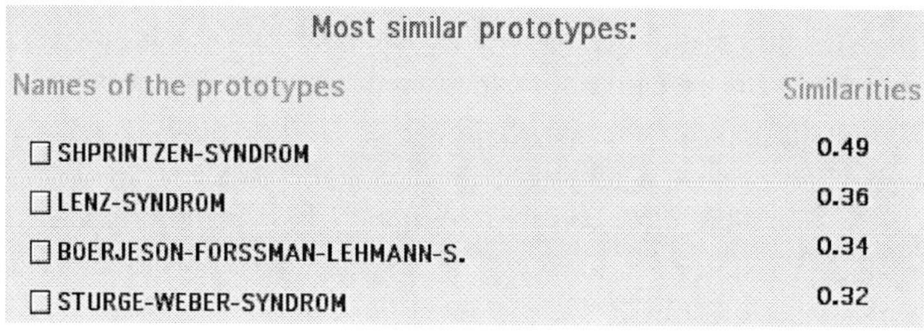

Figure 1. Most similar prototypes after applying a prototypicality measure

2.2. Adaptation Rules

In the fourth and final step, the user can optionally choose to apply adaptation rules on the syndromes. These rules state that specific combinations of symptoms favour or disfavour specific dysmorphic syndromes. Unfortunately, the acquisition of these adaptation rules is very difficult, because they cannot be found in textbooks but have to be defined by experts of paediatric genetics. So far, we have got only 10 of them and so far, it is not possible that a syndrome can be favoured by one adaptation rule and disfavoured by another one at the same time. When we, hopefully, acquire more rules, such a situation should in principle be possible but would indicate some sort of inconsistency of the rule set.

How shall the adaptation rules alter the results? Our first idea was that the adaptation rules should increase or decrease the similarity scores for favoured and disfavoured syndromes. But the question is how. Of course no medical expert can determine values to manipulate the similarities by adaptation rules and any general value for favoured or disfavoured syndromes would be arbitrary. So, instead the result after applying adaptation rules is a menu that contains up to three lists (figure 2). On top the favoured syndromes are depicted, then those neither favoured nor disfavoured, and at the bottom the disfavoured ones.

Additionally, the user can get information about the specific rules that have been applied on a particular syndrome. For the "Lenz-Syndrome, for example, the applied rule no. 6 states:

```
if the patient has medial diffuse hypoplast brows
    and if the patient has prominent corpus-anthelicis
    then the Lenz-Syndrome is very probable
```

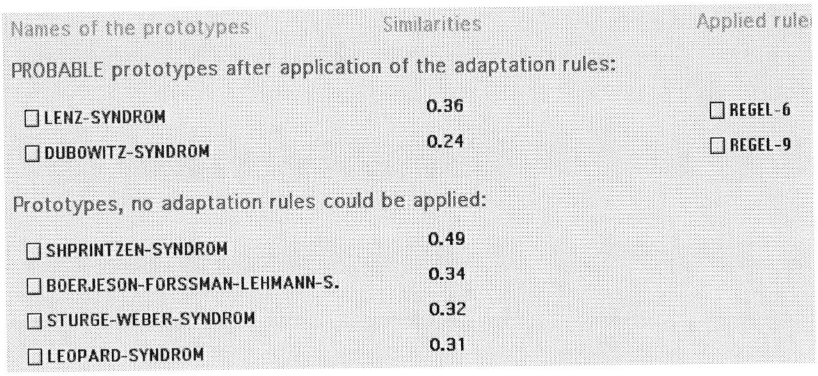

Figure 2. Most similar prototypes after additionally applying adaptation rules

In the example (figures 1 and 2), the correct diagnosis is Lenz-syndrome. The computation of the prototypicality measure of Rosch and Mervis determines Lenz-syndrome as the most similar but one syndrome (here Tversky's measure provides a similar result, only the differences between the similarities are smaller). After application of adaptation rules, the ranking is not obvious. Two syndromes have been favoured, the more similar one is the right one. However, Dubowitz-syndrome is favoured too (by a completely different rule), because a specific combination of symptoms makes it probable, while other observed symptoms indicate a rather low similarity.

3. Results

Cases are difficult to diagnose when patients suffer from a very rare dysmorphic syndrome for which neither detailed information can be found in literature nor many cases are stored in our case base. This makes evaluation difficult. If test cases are randomly chosen, frequently observed cases resp. syndromes are frequently selected and the results will probably be fine, because these syndromes are well-known. However, the main idea of the system is to support diagnosis of rare syndromes. So, we have chosen our test cases randomly but under the condition that every syndrome can be chosen only once.

3.1. Application of Adaptation Rules

For 100 cases we have compared the results obtained by both prototypicality measures (table 1).

The results may seem to be rather poor. However, diagnosis of dysmorphic syndromes is very difficult and usually needs further investigation, because often a couple of syndromes are very similar. The first step is to provide the doctor with information about probable syndromes, so that he gets an idea about which further investigations are appropriate. That means, the right diagnose among the three most probable syndromes is already a good result.

Obviously, the measure of Tversky provides better results, especially when the right syndrome should be on top of the list of probable syndromes. When it should be only among the first three of this list, both measures provide equal results.

Table 1. Comparison of prototypicality measures

Right Syndrome	Rosch and Mervis	Tversky
on Top	29	40
among top 3	57	57
among top 10	76	69

Table 2. Results after applying adaptation rules

Right Syndrome	Rosch and Mervis	Tversky
on Top	32	42
among top 3	59	59
among top 10	77	71

Table 3. Results after applying some more adaptation rules

Right Syndrome	Rosch and Mervis	Tversky
on Top	36	44
among top 3	65	64
among top 10	77	73

3.2. Application of Adaptation Rules

Since the acquisition of adaptation rules is a very difficult and time consuming process, the number of acquired rules is rather limited, namely at first just 10 rules. Furthermore, again holds: the better a syndrome is known, the easier adaptation rules can be generated. So, the improvement mainly depends on the question how many syndromes involved by adaptation rules are among the test set. In our experiment this was the case only for 5 syndromes. Since some had been already diagnosed correctly without adaptation, there was just a small improvement (table 2).

Some more adaptation rules. Later on we acquired eight further adaptation rules and repeated the tests with the same test cases. The new adaptation rules again improved the results (table 3).

It is obvious that with the number of acquired adaptation rules the quality of the program increases too. Unfortunately, the acquisition of these rules is very difficult and especially for very rare syndromes probably nearly impossible.

3.3. Application of Automatically Acquired Adaptation Rules

When the number of adaptation rules increases, an improvement can be observed. However, the medical expert was not able to provide further rules. So, we decided to attempt to generate them automatically. For each syndrome (prototype) the corresponding patients were considered, especially the frequencies of their symptomes. First, the most frequently observed symptomes of all cases belonging to a prototype were determined. To do this a threshold for "most frequent" was used. Subsequently, all combinations of two of these most frequent symtomes were generated and then the frequencies of all such combinations were computed, which should be higher than a second threshold. First, for both thresholds were used very high values, namely 90%, afterwards we decreased them step by step to get more and more possible adaptation rules. With a long list of possible rules (combinations of two symptomes occurring very frequently with a specific syndrome) we consulted the medical expert who decided which of them should be appropriate. So, the adaptation rules were generated automatically but the decision about their appriateness lay by the expert. The expert accepted two groups of rules, namely very probable ones (group A) and rather vague ones (group B). Group A contains just seven rules, wheres group B contains 22 rules. Of course, all automatically generated rules are favouring specific syndromes. With the described generation method no disfavouring rules can be generated.

Illustration of the automatic generation of adaptation rules. For illustration purposes the prototype-j may have just three cases, which have a few symptomes (figure 3). Most of these symptomes occur in two of the three cases. They are selected in the first step. Only the combination "symtome-3 and symtome-12" occurs together in two cases. So, the generated rule states:

```
if the patient has symptome-3
   and if the patient has symptome-12
   then the syndrome of prototype-j is probable
```

These new adaptation rules could improve the results slightly further (tables 4 and 5). However, for beeing among the top 10 nearly no improvement could be observed, which means that newly correctly favoured syndrome had already been among the top 10 before.

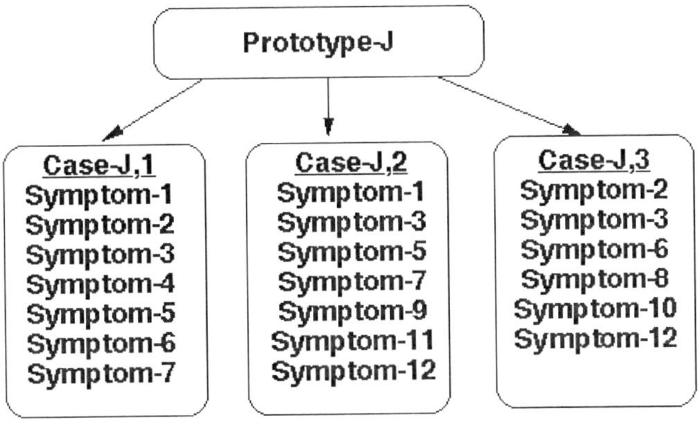

Figure 3. Illustration-Prototype-J with three cases

Table 4. Results after adationally applying seven very probable adaptation rules (group A)

Right Syndrome	Rosch and Mervis	Tversky
on Top	38	45
among top 3	67	67
among top 10	77	73

Table 5. Results after adationally applying 22 vague adaptation rules (group B)

Right Syndrome	Rosch and Mervis	Tversky
on Top	41	46
among top 3	71	70
among top 10	78	73

4. Conclusion

Diagnosis of dysmorphic syndromes is a very difficult task, because many syndromes exist, the syndromes can be described by various symptoms, many rare syndromes are still not well investigated, and from time to time new syndromes are discovered.

We apply a method that does not retrieve single cases (as in CBR) but generalised, prototypical cases. Each of these prototypes represents and characterises one specific diagnosis. Though we have used it just for dismorphic syndromes, we assume that this idea is rather typical for diagnostic tasks, because it seems to be reasonable to search for a general description of a disease instead of searching for single patients.

We have compared two prototypicality measures, where the one by Tversky provides slightly better results. Since the results were rather pure, we additionally have applied adaptation rules. We have shown that these rules can improve the results. Unfortunately, the acquisition of them is very difficult and time consuming. Furthermore, the main problem is to diagnose rare and not well investigated syndromes and for such syndromes it is nearly impossible to acquire adaptation rules.

However, since adaptation rules do not only favour specific syndromes but can be used to disfavour specific syndromes, the chance to diagnose even rare syndromes also increases by the count of disfavouring rules for well-known syndromes.

First, we have acquired some adaptation rules by consulting medical experts, later on we generated further fovouring rules automatically. With the number of acquired rules the results improve, but it is just a very small improvement.

References

[1] Strube, G. & Janetzko, D. (1990). Episodisches Wissen und fallbasiertes Schließen: Aufgaben für die Wissensdiagnostik und die Wissenspsychologie. Schweizerische

Zeitschrift für Psychologie, 49(4), 211-221.

[2] Swanson, D. B., Feltovich, P. J. & Johnson, P. E. (1977). *Psychological Analysis of Physician Expertise: Implications for Design of Decision Support Systems.* In: Shires DB, Wolf H, editors. Proc MEDINFO 77, North-Holland, Amsterdam, 161-164.

[3] Schank, R. C. (1982). *Dynamic Memory*: a theory of learning in computer and people. Cambridge University Press, New York.

[4] Bareiss, R. (1989). *Exemplar-based knowledge acquisition.* Academic Press, San Diego.

[5] Maximini, K., Maximini, R. & Bergmann, R. (2003). *An Investigation of Generalized Cases.* In: K. D. Asley, & D. G. Bridge, editors. Proc ICCBR 2003, Springer, Berlin, 261-275.

[6] Bellazzi, R., Montani, S. & Portinale, L. (1998). Retrieval in a prototype-based case library: a case study in diabetes therapy revision. In: B. Smyth, & P. Cunningham, editors. *Proc EWCBR*, Springer, Berlin Heidelberg, 64-75.

[7] Bichindaritz, I. (1995). Case-based reasoning adaptive to several cognitive tasks. In: A. Aamodt, & M. Veloso, editors. *Case-Based Reasoning Research and Development*, Proc ICCBR-95, Springer, Berlin Heidelberg, 391-400.

[8] Turner, R. (1988). Organizing and Using Schematic Knowledge for Medical Diagnosis. In: J. Kolodner, editor. *Proc of a Workshop on CBR*, Florida, 435-446.

[9] Clancey, W. J. (1985). Heuristic Classification. *Artificial Intelligence*, *27*, 289-350.

[10] Selz, O. (1913). Über die Gesetze des geordneten Denkverlaufs. Stuttgart.

[11] Perner, P. (2004). Are case-based reasoning and dissimilarity-based classification two sides of the same coin? *Journal Engineering Applications of Artificial Intelligence*, *15(2)* 205-216.

[12] Taybi, H. & Lachman, R. S. (1990). *Radiology of Syndromes*, Metabolic Disorders, and Skeletal Dysplasia. Year Book Medical Publishers, Chicago.

[13] Gierl, L. & Stengel-Rutkowski, S. (1994). Integrating Consultation and Semi-automatic Knowledge Acquisition in a Prototype-based Architecture: Experiences with Dysmorphic Syndromes. *Artificial Intelligence in Medicine*, *6*, 29-49.

[14] Clinical Dysmorphology. htp://www.clyndysmorphol.com (last accessed: April 2009)

[15] Winter, R. M., Baraitser, M. & Douglas, J. M. (1984). A computerised data base for the diagnosis of rare dysmorphic syndromes. *Journal of medical genetics*, *21(2)* 121-123.

[16] Stromme P. (1991). The diagnosis of syndromes by use of a dysmorphology database. *Acta Paeditr Scand*, *80(1)*, 106-109.

[17] Weiner, F. & Anneren., G. (1989). PC-based system for classifying dysmorphic syndromes in children. *Computer Methods and Programs in Biomedicine*, *28*, 111-117.

[18] Evans, C. D. (1995). A case-based assistant for diagnosis and analysis of dysmorphic syndromes. *International Journal of Medical Informatics*, *20*, 121-131.

[19] Tversky, A. (1977). Features of Similarity. *Psychological Review*, *84(4)*, 327-352.

[20] Kolodner, J. (1993). *Case-Based Reasoning*. Morgan Kaufmann Publishers, San Mateo.

[21] Aamodt, A. & Plaza, E. (1994). Case-Based Reasoning: Foundation issues, methodological variation, and system approaches. *AICOM*, *7*, 39-59.

[22] Broder, A. (1990). Strategies for efficient incremental nearest neighbor search. *Pattern Recognition*, *23*, 171-178.

[23] Wilke, W., Smyth, B. & Cunningham, P. (1998). Using configuration techniques for adaptation. In: M., Lenz, B., Bartsch-Spörl, H. D. Burkhard, & S. Wess, editors: Case-Based Reasoning technology, from foundations to applications. *Springer-Verlag*, Berlin Heidelberg New York 139-168.

[24] Rosch, E. & Mervis, C. B. (1975). Family Resemblance: Studies in the Internal Structures of Categories. *Cognitive Psychology*, *7*, 573-605.

In: Case-Based Reasoning: Processes Suitability... ISBN: 978-1-61728-352-9
Editors: Antonia M. Leeland

Chapter 8

NEW APPROACH OF CASE-BASED REASONING*

***Chokri El Aoun*[†a], *Hichem Eleuch*[b],**
***Hella Ben Ayed*[a] *and Esma Aïmeur*[c]**

[a] Laboratoire CRISTAL, Ecole Nationale des Sciences de l'Informatique, Université de Manouba, Tunis, Tunisie

[b] Centre National des Sciences et Technologies Nucléaires, Tunis, Tunisie. Associate member in the Abdus Salam International Centre of Theoretical Physics (ICTP), Trieste, Italy

[c] Département d'informatique et de recherche opérationnelle, Université de Montréal. Pavillon André Aisenstadt, Montréal (QC), Canada

ABSTRACT

Case-Based Reasoning (CBR) allows us to resolve problems in a dynamic environment, and propose a solution that follows a checking step, in which we proceed along the test-error-correction cycle until we reach the result aspired to. This paper proposes a novel model for CBR (the 3R model), in which Retrieve, Reuse, and Retain are the main tasks for the CBR process. The integration of mathematical reasoning allows the search for the weights that are assigned to the different attributes of a case, then, to come up with the wished for result, based on the similar case in one go. Hence, the classical 4R model is reduced through the elision of the Revise step, the results are then reached automatically (the search of the weights, the similar case and the final result). This model is used as a negotiation strategy to predict the seller's behaviour. We have applied it to the real estate domain and we have come up with interesting results.

Keywords: Case-based reasoning, 3R model, similarity, prediction, negotiation.

---

* A version of this chapter was also published in Chaos and Complexity Letters, Volume 4, Issue 1, edited by Franco Orsucci and Nicoletta Sala, Nova Science Publishers. It was submitted for appropriate modifications in an effort to encourage wider dissemination of research.

† Corresponding author: chokri.elaoun@laposte.net, Address: IHEC-Sfax B.P. 1170 – Sfax 3018 – Tunisie.

1. INTRODUCTION

The Case-Based Reasoning (CBR) allows the use of some specific knowledge of previously acquired experiences to solve new problems [1, 2, 4, 16, 20, 25]. The utmost goal of using acquired experience is to improve a system's performances. Actually, by relying on experience we are able to: devise a kind of reasoning short cut, avoid reproducing previously made mistakes and make knowledge acquisition easier. Therefore, the major concern here is efficiency of the system and its gradual improvement both at the level of its completion and efficiency.

The very basic idea of CBR is to solve new problems by comparing them to problems already solved [1, 4, 20]. Traditional case-based reasoning is based on the assumption that similar problems have similar solutions. Hence, during retrieval the most similar case or the most similar cases in the case base are selected. Then, during reuse the information in the retrieved case(s) is used to solve the new problem. The new problem description is combined with the information contained in the old case to form a solved case. The result of the reuse phase is a solved case that is suggested to the user. During revision the applicability of the proposed solution is evaluated in the real world. If necessary and possible, the proposed case must be adapted manually in some way. If the case solution generated during the revise phase must be kept for future problem solving, the case base is updated with a new learned case in the retain phase [24].

Increasing attention has been drawn to the CBR in developing Web-based intelligent systems. Several Web-based systems that use CBR are already in existence [11]. The traditional CBR-cycle [1] can not be directly applied to the process of electronic commerce where the consumer has a larger space and a wider range of products and solutions to choose from. The consumer is able to personalize the desired product and to modify his request until his requirements are met [31]. The way to proceed in an Electronic Commerce (EC) context is different from the traditional commerce context where the consumer buys what he sees.

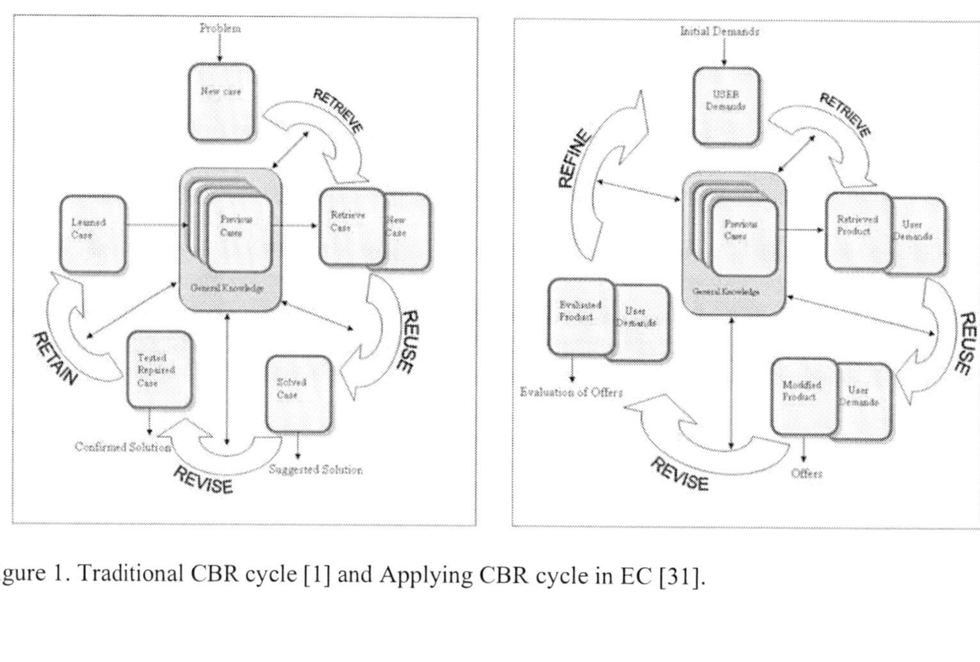

Figure 1. Traditional CBR cycle [1] and Applying CBR cycle in EC [31].

The use of the CBR, which is a tool for problem solving, brings in an added value if integrated in the devices of EC namely the online sales supports, negotiation, products recommendation or online technical assistance. Thus, we will end up with smart supports. These observations make the use of the CBR technique more plausible and the prediction of a seller's behaviour possible through a model. In this research, our approach allows the understanding of the seller's strategy; so that we are able to take the right decision. This is a new CBR approach because it rests on a 3-phase cycle: research, reuse and retain, hence the label 3R model. In order to demonstrate its efficiency we have tested it in the field of real estate negotiation.

The remainder of this paper is structured as follows:

Section 2 presents negotiation and some strategies of various technologies employed in negotiation to generate offers or counter-offers. In section 3, we present our approach and its use in real estate negotiation. Section 4 validates the 3R model through tests in the fields of real estate. Eventually, in section 5 we come up with conclusive remarks and a few research perspectives

2. NEGOTIATION

Negotiation is a resource allocation mechanism and a decision making process through which participants (buyer and seller) make iterative exchanges with decentralized manner via offers and counter-offers with the goal of maximizing their interests.

Negotiations conducted over the Web are commonly called *e-negotiations* which use *e-negotiation systems* (ENSs). E-negotiation is a process that involves people and ENSs. While passive systems can be seen as fast and sophisticated messengers, active systems can facilitate, support and mediate. The systems that can access and process knowledge and are able to work independently of their users, that is, they have a certain degree of intelligence and can be proactive [15, 33].

During a negotiation process, a negotiator may consider questions such as [18]: What should be my bottom line? What is a reasonable expectation? On which issues should I remain firm and on which should I be more flexible? How rapidly should I be willing to make concessions [12]? Answers to these questions will shape a negotiator's strategy [6]. In other words, negotiation behavior is often described in terms of different strategies [23], and it is thought that a negotiator's strategy can be determined from his negotiation behavior. Many studies have investigated negotiation strategies [6, 9, 10, 12, 19, 29], but most have been theoretical or based on data obtained only from questionnaires.

Traditionally, attempts to understand different aspects of negotiations have used many perspectives, such as game theory, psychology, political science, communication, labor relations, law, sociology, anthropology, operation researchers, and artificial intelligence [18].

CBR approach is extensively used for negotiation [17]. The CBR-based negotiation system aims at identifying the demands in cooperation with the customers and finding a product that fulfils them [35]. During negotiation in e-sales, the CBR-based negotiation system might suggest or even add some new demands or modify some weak demands for the purpose of finding an appropriate product. For configurable products, it is also possible for the CBR-based negotiation system to modify existing products during product (solution) adaptation to meet the customer's demands.

Zhang [34] presented a web-based negotiation agent that applies CBR techniques to capture and reuse previously successful negotiation cases. The case-based negotiator generates a proper concession for negotiation episode based on retrieval, selection and reuse of relevant previous experience [8]. In this scenario, the offer of the opponent is first assessed to see if it is acceptable. If it is not, a strategy that would entail a counter offer is conceived. In a case base, a number of previous negotiation cases called experiences are stored. Then, certain CBR techniques are applied to reuse these experiences, i.e., the strategy of a previous negotiation case is used to set the current negotiation strategy. Hence, as a first stage, the appropriate experiences are retrieved. Then, the strategy that corresponds most is selected. Finally, this strategy is employed in the current negotiation to generate a counter offer.

Agents are the computational entities that participate in the negotiation process. Each agent is assumed to be capable of rating its preferences, so that it can evaluate and choose between different deals. But there is no right strategy to generate a good proposition, because most current e-commerce systems use predefined and non-adaptive negotiation strategies in the generation of offers and counteroffers during the course of the negotiation [32].

Chavez and Maes have created Kasbah, a marketplace for negotiating the purchase and sale of goods using intelligent software agents. The agents are, in their words, "not tremendously smart," nor do the agents use any machine learning or AI techniques, nor do the agents attempt to encompass abstractions such as user goals or preferences. Rather, the Kasbah software agents receive their complete strategies through a World Wide Web form the users, who specify strategy and the way in which the acceptable price can change over time, and who retain final control over the agents at all times. Chavez and Maes report the user feedback was generally positive, but the participants were disappointed when their agents did "clearly stupid things," such as accepting the first feasible offer when a second, better one was available [5]. With eBay, for instance, the consumers manage their own negotiation strategies over an extended period of time [21]. In the AuctionBot [13], the users specify the names of the auctions they want to participate in, the initial amounts of the goods, and the bidding strategies they prefer.

The approach by Su [26] is based on the idea that a consumer registers on a proxy negotiation server by giving a description of the goods or services she wants, her preferences, and a negotiation strategy. The server uses the negotiation strategy supplied by the consumer. The user will have no control over the negotiations once they are started. He can only see their progress, but cannot intervene in them. Another approach is that of genetic programming, which uses techniques akin to Darwinian evolution to select winning negotiation strategies from a large population of initial possibilities [22]. The major apparent disadvantage of genetic programming is it requires many trials to achieve the good strategies in the end. This number varies from about 20 generations [22] to upwards of 4000 generations [7] and all runs must be made against opponent(s) which are as realistic as possible. Hence, it may be unrealistic to "teach" a genetic algorithm using a human opponent because of time constraints.

Broadly speaking, what these models lack is their inability to product the selling price. In order to remedy these difficulties we suggest a model that makes the anticipation of the seller's behavior possible. We prefer the user to be in control and intervene whenever he wants in the negotiation.

3. Description of Our Approach

Our approach is set in a dynamic environment. This way, we propose a negotiation strategy that enables us to predict the selling price of a good, even if we are confronted with a new situation (a new good).

In order to facilitate the problem solving and the retrieval of the pertinent solution, we make use of CBR technology. The basic idea of CBR is to solve new problems by comparing them to problems already solved [4, 20]. The key assumption is that if two problems are similar, then their solutions are similar too. Our approach is based on the CBR, a technique that adapts well to prediction problems. In fact, it could be used when we have little information about the problem to be solved and for which an optimal solution is primarily unknown.

In what follows we suggest our approach inspired from the classical cycle [1]. This approach is based on a three-stage cycle (retrieve, reuse and retain). Then we break the 3R cycle applying it to real estate negotiation. This model enables us to predict the selling price in the course of a negotiation, to adopt a good strategy and a good decision-making tool. It also allows us to assess in order to understand the behaviour of the seller and to predict in order to monitor his strategy. This approach is novel as far as the other negotiation approaches mentioned previously are concerned (section 2).

3.1. The 3R Model

The 3R model is based on the CBR technology that is a method for experience-based problem solving: new problems are solved based on stored experience about similar previously solved problems. Therefore, previous problems and related solutions are stored in a case base. When a new problem must be solved, an optimal weights of a problem attributes are determined in first, next, a similar old problem is searched in the case base and then the solution connected with this problem is used to calculate the new solution and so to solve the new problem. The 3R model is based on a three-stage cycle (Figure 2) inspired from the traditional CBR cycle [1].

Retrieve

This stage takes place in two phases (retrieve 1 and retrieve 2). The first phase of this stage is "retrieve 1", it has as objective to look for the optimal weights. This process of research is iterative, in which an adjustment of weights is carried out through an increment to get optimal weights. Then, a second phase, retrieve 2, enables us to draw the similar case once the appropriate weights are found.

Reuse

This stage allows working out the solution of the target case drawing on the solution of the similar case. In this stage, to reuse is not to copy or modify the similar solution to reach the final solution, but it is to apply a formula to generate the final target solution. Therefore, there is no iterative process of modification, checking or correction of the similar solution.

Retain

The last stage as in the traditional cycle is the integration or the storing of the new case together with its solution, and the updating of the case-base.

3.2. Real Estate Negotiation According to the 3R Model

Leake explains that the world is regular: similar problems have similar solutions [20]. Consequently, solutions for similar prior problems are a useful starting point for new problem solving. The paradigm of the basis of the reasoning by analogy consists in the setting of correspondence between two situations -sometimes belonging to two different fields-, according to memorized information on already met situations, in order to provide an adequate behavior facing a new situation [12, 13, 18].

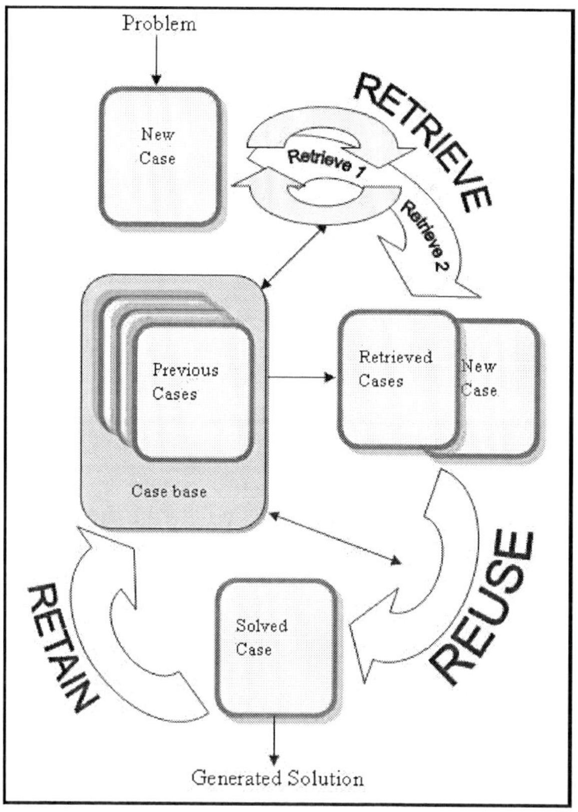

Figure 2. The 3R model to anticipate the seller's behavior.

In order to prove the validity of our approach, we have chosen real estate negotiation as a working example. In a real estate agency there is a set of properties that need to be sold. In this domain there are two main players: seller and buyer. The seller agent acts on behalf of the interests of the real estate agency, while the buyer agent represents his own interests. This is an obvious conflict of interest that is usually resolved by a negotiation.

If the seller know the potential buyers and their valuations of the property for sale, his pricing problem would have a simple solution, and thus, he can predict their behaviors. The trouble is the buyer usually has only incomplete information about sellers' valuations. Therefore, the buyer has to figure out a pricing scheme that performs well even under incomplete information, and therefore, he can make his own estimate of the value of the property. A basic ingredient of the negotiation process is the correct anticipation of the other side's actions. An adequate offer reflects an exact prediction of the object's value and the best choice of a strategy leading to maximize self interest.

In order to make good decisions, there is a need to look for more efficient and rapid strategies [14]. Therefore, the seller's behaviour has to be countered and monitored through the anticipation of the selling price of the property. As such, the efficiency of our research is found in the hold information of utmost importance about the property in negotiation and the price the seller wants to get. In order to conceive this strategy, we apply our 3R model.

There is an experience principle in business activities, for example, "Two properties with similar features have similar values". The underlying idea of CBR is simple: Do not solve problems from scratch but remember how you (or someone else) solved a similar problem and apply this knowledge to solve your current problem [3]. For case-based problem solving, a set of cases is the primary source of knowledge. Cases are representations of previous experience. Each case consists at least of a problem description and a related solution to the problem or some kind of information that is relevant to determine the solution [24]. This resolute problem is stored in a case base [3]. In the next, we define case and case base.

Case

The smallest experience item in CBR is called a case. When applying the structural CBR approach, each case is described by a finite and structured set of attribute-value pairs that characterize the problem and the solution. Hence, a single case can be considered as a point in the space defined by the Cartesian product of the problem space **P** and solution space **S** [28].

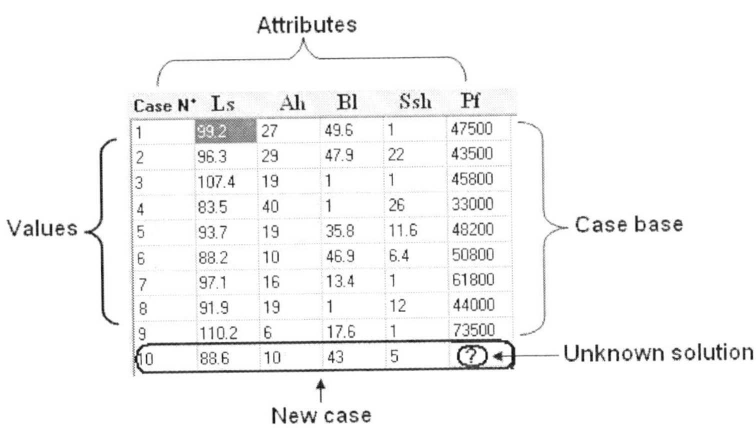

Case N°	Ls	Ah	Bl	Ssh	Pf
1	99.2	27	49.6	1	47500
2	96.3	29	47.9	22	43500
3	107.4	19	1	1	45800
4	83.5	40	1	26	33000
5	93.7	19	35.8	11.6	48200
6	88.2	10	46.9	6.4	50800
7	97.1	16	13.4	1	61800
8	91.9	19	1	12	44000
9	110.2	6	17.6	1	73500
10	88.6	10	43	5	?

Figure 3. Attraction Case base and repulsion of charges

A case is therefore the association of a problem and its solution,

C = (P, S) where:
P: is the problem, it is an element of the space of the **P** problem.
S: is the solution that is an element of the space of the **S** solution.

A source case is a case that is going to inspire us to resolve a new case that we will call the target case. A source case is written: $c_s = (p_s, s_s)$ and a target case is written: $c_T = (p_T, s_T)$.

Case base

A case base is a collection of cases of resolution of the same problem. As for our base of dealt with cases of property sales, we have for each case a description of an episode of a sale resolution and its describers that define the problem associated to its solution. The first five columns correspond to the describers of the problem (Figure 3) and the final one "Pf" corresponds to the describer of the solution (in this case, only the selling price has been taken into consideration).

3.3. The 3R Model Cycle

The following flowchart illustrates the different steps of the 3R model cycle without taking into consideration the "retain" stage. Then, in the following section the whole process will be minutely presented.

3.3.1. Retrieve

We try to predict the closing price (final price or selling price) of the share, knowing the opening price (starting price). Our study takes into account the fact that the transaction volume, the spread, the return and the volatility influence the behavior of a decider in the stock exchange and particularly in the Tunis Stock Exchange.

In order to make the comparison of the cases possible, we have to be able to compare their attributes' values so as to see to what extent these are similar.

The working-out of a new case consists in facilitating the description of the problem so as to look for a case, the solution of which would be the most easily adaptable. The adaptability of a case amounts to the adaptation "effort" that would be necessary for the past solution (of the source case) to serve as the basis for the current solution (of the target case). The common method is to complete or filter the description of a problem relying on the knowledge of the field in question, to infer whatever possible from an incomplete description, and ponder the describers according to the identified correspondences between the describers of the target problem and those of the sought solution.

General Flowchart

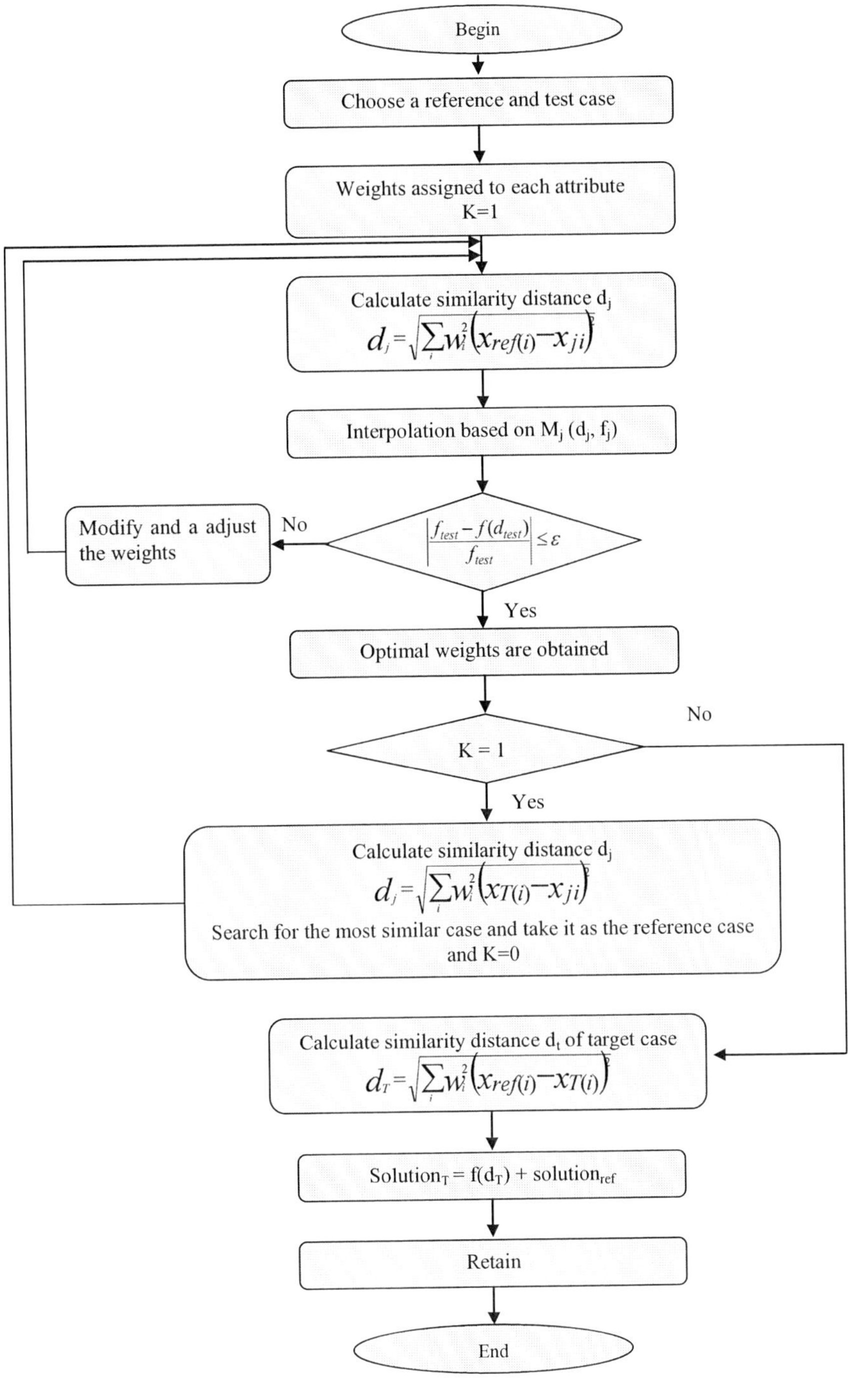

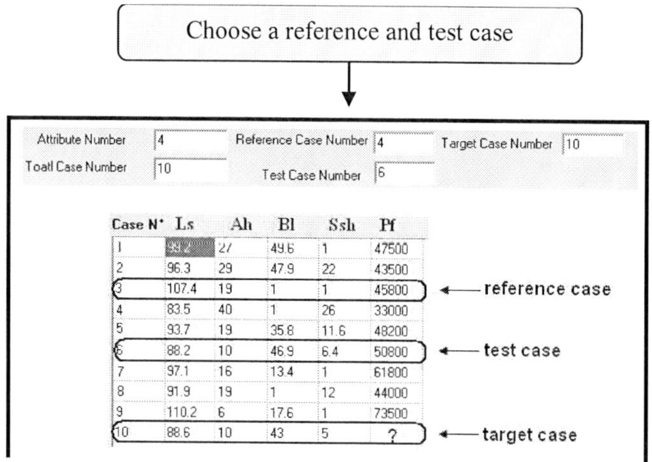

Figure 4. Cases identification

During this stage we go through two phases:

- retrieve 1: it enables the search for the optimal weights
- retrieve 2: it makes it possible for the similar case to be found

A) Retrieve 1: Optimal weights search

In retrieve 1, we begin by identifying the reference and test cases (Figure 4), and then we move on to the specification of the optimal weights.

B) Optimal weights specification

It is difficult to reach the optimal weights from the first choice. For this reason we proceed this way in this step:

a-1) Initialization of the weights
a-2) Calculation of the similarity distance
a-3) Adjustment of the weights
a-4) Calculation of the optimal weights

B-1) Initialization of the weights

Weight values are assigned in an intuitive way so that the increment or the modification of the weights allows us to cover the highest number of possibilities to reach optimality. To carry out this operation, we have to start by assigning a low value to weight "W_1", then to weight "W_2" a value that is less low, and we assign the other value to the remaining weight. During this modification the weights increments operate like a meter.

W_i: Weights assigned to each attribute according to a specific degree of importance
i: attribute number.
K=1: Boolean variable

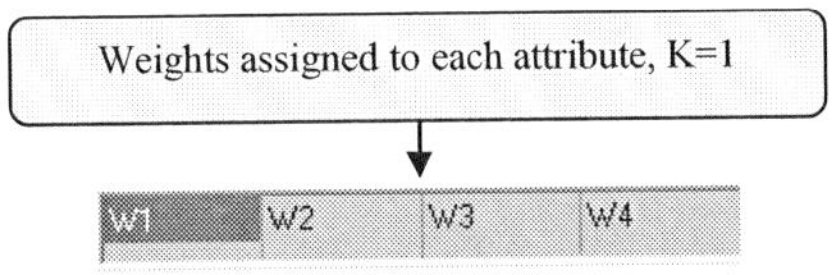

Once we have introduced weights, the reference case and the base cases, we can calculate the similarity distance and this is what the following is about.

B-2) Calculation of the similarity distance

In this step we are going to look for the most similar case to the reference case we have already chosen. The search for the source (similar) case is naturally fundamental in the cycle.

It is worth reminding that the source case to be chosen is normally the case that has the closest problem description ever possible to the description of the target problem, in the most available category of solutions.

It is necessary to define a similarity measurement that would take into account the influence of any variation of the value of the problem attribute on any variation at the level of the value of a solution attribute. Intuitively, we see that problem attributes that have significant bearings on the solution should be given an important "weight", whereas a small value should be given to those having little influence on the solution.

However, metrics and in particular similarity metrics play an important role in case retrieval in CBR. Usually similarity metrics are used to evaluate the similarity between two cases in CBR [27]. In what follows, the section will examine the theoretical similarity metrics based on local similarity and global similarity.

Let p_1 and p_2 be two problems in the possible space of problem **P**, in which every problem has n (feature) attributes. The attribute-value representation of the problem in **P** can be taken as a n-tuple vector; that is:

$p_1 = \{x_{11},\ldots, x_{n1}\}$
$p_2 = \{x_{12},\ldots, x_{n2}\}$

For every $i \in \{1,\ldots,n\}$, there is a similarity S_{mi} on the domain of attribute A_i; that is, S_{mi}: $A_i \rightarrow R$, S_{mi} is called a local similarity metric, and $S_{mi}(x_{i1}, x_{i2})$ is the similarity degree between x_{i1} and x_{i2}.

Local similarity deals with the values of an individual attribute or feature of a problem. However, a problem/solution description has a number of attributes in a CBR system. Therefore how to get an overall similarity assessment for a problem/solution description based on the local similarity assessment is an important part in CBR. The evaluation of global similarity between two multiple-feature descriptions is obtained by aggregating the evaluation of local similarities for each feature. In what follows, the section will examine the theoretical global similarity based on local similarity discussed above from the viewpoint of CBR.

Let g be a composite function from Rx…R to R. Then the global similarity degree of p_1 and, p_2, $S_g(p_1,p_2)$, can generally be considered as

$$S_g(p_1,p_2)=g(S_{m_1}(x_{11},x_{12}),\ldots,S_{m_n}(x_{n1},x_{n2})) \quad (1)$$

Where S_g is a global similarity metric. If is a linear function such that

$$S_g(p_1, p_2) = \sum_1^n w_i S_{m_i}(x_{i1}, x_{i2}) \quad (2)$$

Where W_i is the weighted value of attribute A_i, which reflects the relative importance of corresponding A_i, within the problem in **P** and satisfies $w_i \in [0,1]$ *and* $\sum_1^n w_i = 1$ (normalized weights), (2) is called the weighted Hamming similarity metrics, because its form is essentially the same as the weighted Hamming distance.

Another popular (weighted) global similarity metric is the Euclidean similarity as follows:

$$S_g(p_1, p_2) = \sqrt{\sum w_i^2 \cdot S_{m_i}^2(x_{i1}, x_{i2})}, \quad (3)$$

owing to that its form is essentially the same as the traditional Euclidean distance.

In our situation the resemblance is measured with similarity distance (pondered Euclidean distance) [30]. This distance is defined as d_j which is the distance of case j, in relation to the reference case (Figure 5).

The objective is the use of optimal weights to calculate distance d_j:

$$d_j = \sqrt{\sum_i w_i^2 (x_{ref(i)} - x_{ji})^2} \quad (4)$$

d_j: the similarity distance of case j in relation to the reference case
W_i: Weights assigned to each attribute according to a specific degree of importance
i: attribute number
j: Case number
$x_{ref(i)}$: The attribute i of the reference
x_{ji}: attribute i of the case j

Example: Let's look for the similarity distance "d_4" that means the Euclidean distance between the case references (case 3) and the case 4. We get:

$$d_4 = \sqrt{0.1^2 * (107.4 - 83.5)^2 + 0.2^2 (19 - 40)^2 + 0.3^2 * (1 - 1)^2 + 0.4^2 (1 - 26)^2}$$

$$d_4 = 11.1063991$$

The similarity distance for each case is:

Table 1. Calculation of d_i and f_j

i	d_i	f_i
1	14.6904323	1 700.00
2	16.5456036	-2 300.00
4	11.1063991	-12 800.00
5	11.3511277	2 400.00
7	3.9063154	16 000.00
8	4.66502947	-1 800.00
9	5.62483778	27 700.00

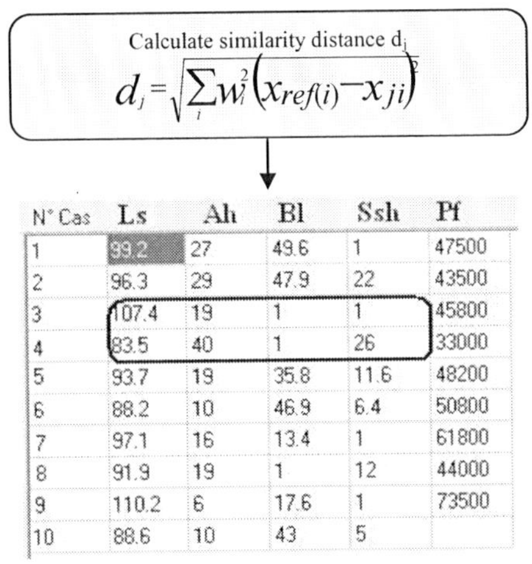

N° Cas	Ls	Ah	Bl	Ssh	Pf
1	99.2	27	49.6	1	47500
2	96.3	29	47.9	22	43500
3	107.4	19	1	1	45800
4	83.5	40	1	26	33000
5	93.7	19	35.8	11.6	48200
6	88.2	10	46.9	6.4	50800
7	97.1	16	13.4	1	61800
8	91.9	19	1	12	44000
9	110.2	6	17.6	1	73500
10	88.6	10	43	5	

Figure 5. Similarity distance

Our aim is to find optimal weights crucial for the calculation of the similarity distance in relation to the target case. The following section deals with this issue.

B-3) Adjustment of the weights

In this section we will be looking for the right weights, an adjustment has to be made if these are not reached. Then we move on to the drawing of the similarity graph, then to the interpolation for the optimality test.

For each case, we assign points to make a similarity diagram. This way, to a given case (case j) corresponds a point made up of a pair d_j and f_j marked as M_j (d_j, f_j). Each point M_j is characterized by its similarity distance d_j and a deviation f_j, which is the disparity between the solution_j and the reference case solution (Figure 6).

Interpolation based on M_j (d_j, f_j)

↓

From the available cases, we draw the interpolated curve and we come up with:

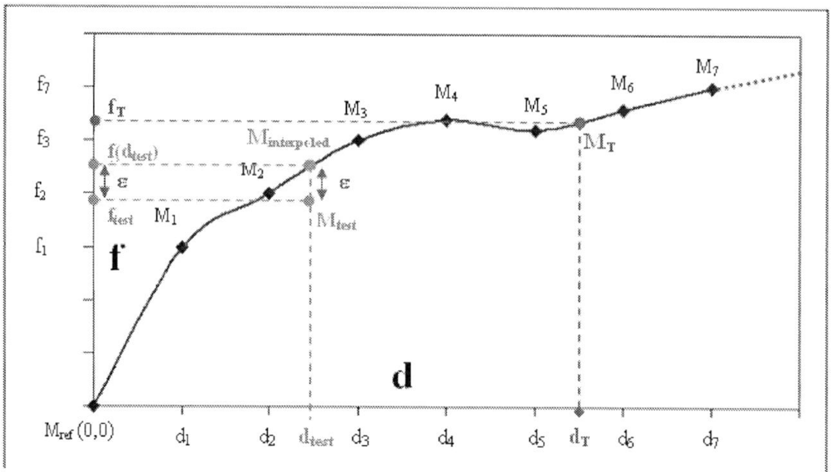

Figure 6. Interpolation curve

M_j: the M_j point assigned to case j

- d_j: the similarity distance of case j in relation to the reference case
- f_j: it is the deviation in relation to the reference solution, i.e., the difference between the solution of case j ($solution_j$) and the reference solution ($solution_{ref}$),

$$f_j = solution_j - solution_{ref} \tag{5}$$

M_{test} (d_{test}, f_{test}) is the point composed of

- d_{test}: the similarity distance of the test case in relation to the reference case.
- f_{test}: it is the difference between the test solution ($solution_{test}$) and the reference solution ($solution_{ref}$),

$f_{test} = solution_{test} - solution_{ref}$ (based on (5))
$M_{interpoled}$ (d_{test}, $f(d_{test})$) is the point composed of

- d_{test}: the similarity distance of the test case in relation to the reference case.
- $f(d_{test})$: calculated with interpolation.

Therefore, we have a number of points at our disposal {Mj (dj, fj), j = 0...n}. What is at stake here is to be able to determine or rather to assess the value of "f" in a given abscissa point

x = d of the interval [d_1, dn]. From the already known data, we are to predict $f(d_{test})$ of the point $M_{interpoled}$ (d_{test}, $f(d_{test})$). To solve this problem we use the interpolation technique.

If we come up with the optimal weights, f_{test} has to be approximate to $f(d_{test})$, otherwise we have to adjust the weights and start the search process for the optimal weights again (Figure 7).

$f(d_{test}) = 67618.063698734$

$f_{test} = solution_{test} - solution_{ref}$
$f_{test} = 50800 - 48200 = 2600$
$f_{test} = 2600$

f_{test} is very different from $f(d_{test})$, so we have to modify the weights and then make variations on $f(d_{test})$ from interpolation until we reach an $f(d_{test})$ close to f_{test}, what allows us to obtain the optimal weights.

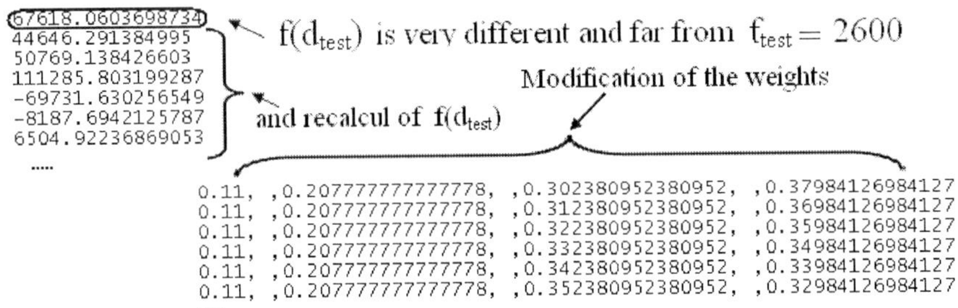

Figure 7. Computation and variation of weights and calculation of f(dtest)

B-4) Calculation of the optimal weights

This step allows the optimal weights to be reached:

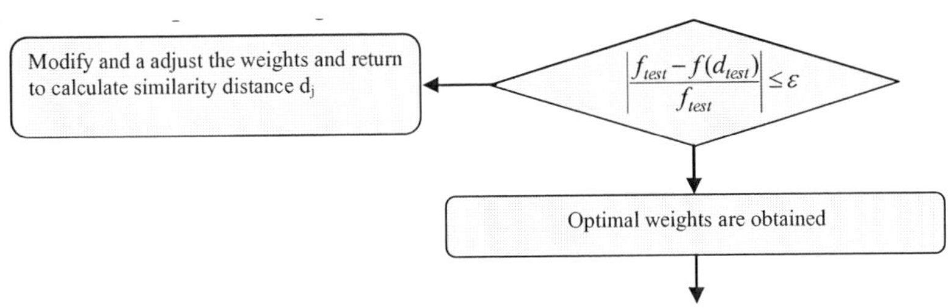

From the test case we have M_{test} (d_{test}, f_{test}) and from the interpolation we get $f(d_{test})$ which is the function f: d → f(d). $f(d_{test})$ is thus a value obtained by interpolation, whereas f_{test} is a value computed according to the difference between case j solution and the reference case solution.

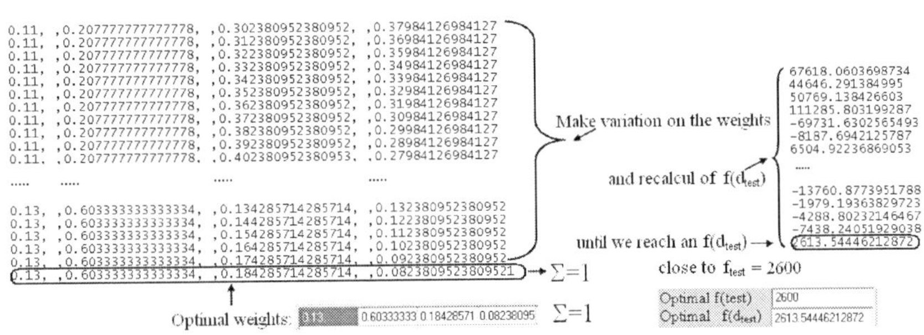

Figure 8. Computation of optimal weights and f(dtest)

If the weights are correct or optimal then we get an f_{test} that tends towards $f(d_{test})$, otherwise we will have to correct and make variations on the weights (Figure 8), so that

$$\left|\frac{f_{test}-f(d_{test})}{f_{test}}\right| \leq \varepsilon \qquad (6)$$

and this way, we reach the optimal weights.

B) Retrieve 2: The search for the similar case

A) Computation of the Similarity Distance in Relation to the Target

Once we have found the optimal weights (W_i), we calculate the similarity distance, i.e. the closest case (the most similar) to the target case, based on (4):

$$d_j = \sqrt{\sum_i w_i^2 \left(x_{cible(i)} - x_{ji}\right)^2}$$

W_i: Optimal weights.
i: Attribute number
j: Case number
$x_{T(i)}$: Attribute i of target case
x_{ji}: Attribute i of case j

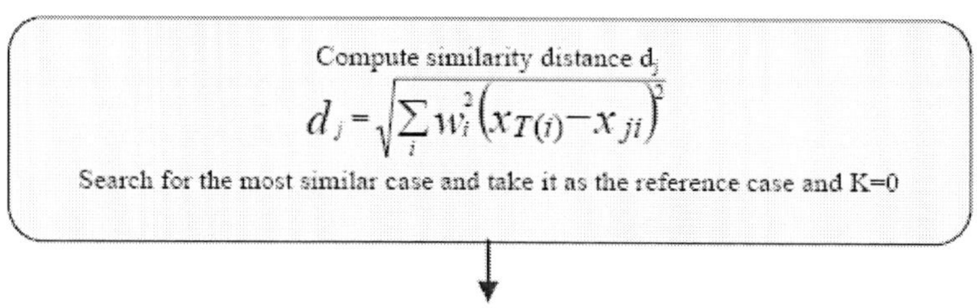

To search for the most similar case (the test case not being taken into consideration) is to ultimately find the distance that is closest to the target case (Figure 9).

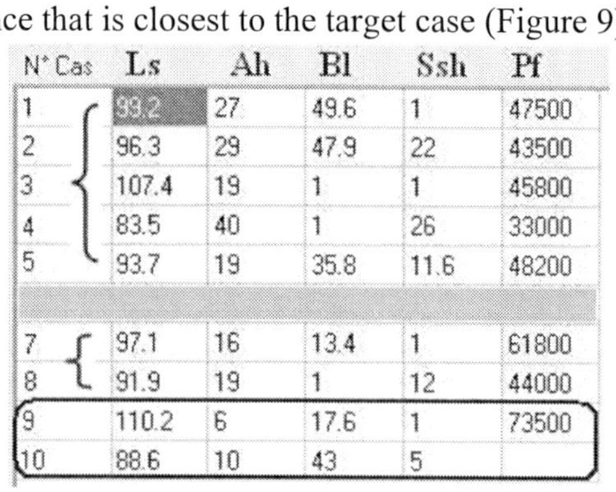

N° Cas	Ls	Ah	Bl	Ssh	Pf
1	99.2	27	49.6	1	47500
2	96.3	29	47.9	22	43500
3	107.4	19	1	1	45800
4	83.5	40	1	26	33000
5	93.7	19	35.8	11.6	48200
7	97.1	16	13.4	1	61800
8	91.9	19	1	12	44000
9	110.2	6	17.6	1	73500
10	88.6	10	43	5	

Figure 9. Computation of the similarity distance

Afterwards, we take this case which the most similar as the most adequate reference case and we take up the same steps again: calculation of the similarity distance, adjustment of the weights and calculation of the optimal weights.

B) Similar Case Retrieval

By the time we have determined the optimal weights, we have calculated the optimal weights of the unknown case (the target case), then we calculate its similarity distance in relation to the reference (Figure 10), based on (4):

$$d_T = \sqrt{\sum_i p_i^2 \left(x_{ref(i)} - x_{T(i)}\right)^2}$$

W_i: Optimal weights.
i: Attribute number
j: Case number
$x_{ref(i)}$: Attribute i of reference case
$x_{T(i)}$: Attribute i of target case

$$d_T = \sqrt{0.13^2 * (99.7-88.6)^2 + 0.6033333^2 * (19-10)^2 + 0.1842857^2 * (35.8-43)^2 + 0.0823809^2 * (11.6-5)^2} = 5.6551431$$

3.3.2. *Reuse*

By computing what the solution to the target case problem would be, inspired by the solution of the most similar source case the adaptation thus ends the "analogical inference".

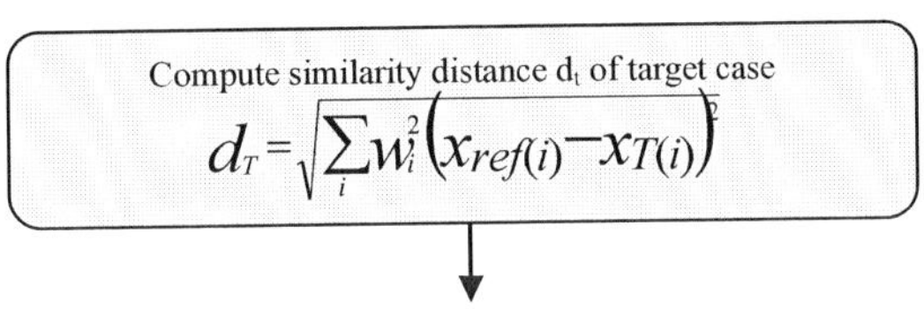

N° Cas	Ls	Ah	Bl	Ssh	Pf	
1	99.2	27	49.6	1	47500	
2	96.3	29	47.9	22	43500	
3	107.4	19	1	1	45800	
4	83.5	40	1	26	33000	
5	93.7	19	35.8	11.6	48200	← Last reference case
6	88.2	10	46.9	6.4	50800	
7	97.1	16	13.4	1	61800	
8	91.9	19	1	12	44000	
9	110.2	6	17.6	1	73500	
10	88.6	10	43	5	49300	← Target case

Figure 10. Computation of the similarity distance and target case

The target solution, which is the final price to be predicted, is reached by (Figure 11), based on (5):

$$Solution_T = f(d_T) + solution_{ref}$$

We get $f(d_T)$ from the interpolation which is in our case $f(d_{10})$.

$$f(d_T) = f(d_{10}) = 985.2423458$$

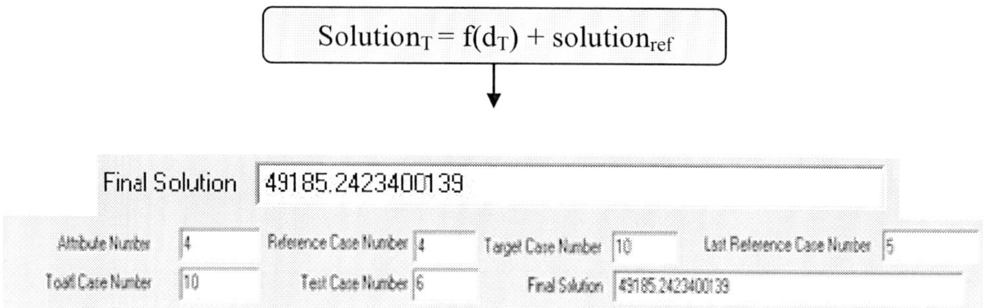

Figure 11. Target Solution

Final Solution = $Solution_T = f(d_{10}) + solution_{ref(5)}$ ⟶ (last reference)

$$Solution_T = 985.2423458 + 48200 = 49185.2423$$

We note that our estimated solution is deviated with 0.23% from the reel solution (49300).

3.3.3. Retain

The resolved case would therefore be retained and stored in the base for later use. The base would be up-dated and this way an incremental modification of the base is done. This allows gradual learning. In fact, the cases are sequentially stored in a base; it is the simplest structure for the memory. The advantage of such a structure is that during the search phase, all the available cases in the memory are tested. This guarantees a sharp search. Furthermore the search is not expensive, it would be enough to add the new case at the end of the file.

The major drawback is that given the size of the memory the time of research gets linearly longer.

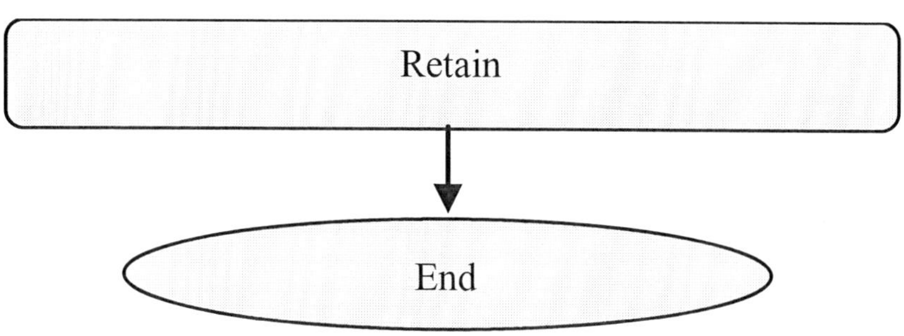

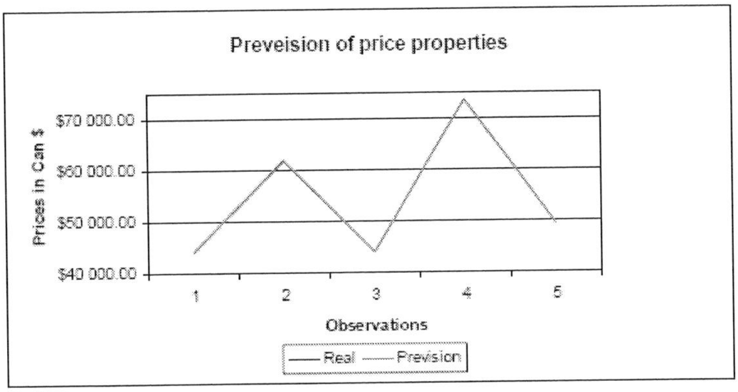

Figure 12. Comparison between real and estimation estate

Table 2. Real estate estimation using 3R model

Prediction from 16 to 20	
Réel	Prévision
$44 000.00	$44 201.62
$61 800.00	$61 952.65
$44 000.00	$44 001.81
$73 500.00	$73 460.98
$49 300.00	$49 405.88

4. Model Validation

To demonstrate the usefulness, the feasibility and the relevance of our approach, we will test and present the experimental results obtained by applying the 3R model to the real estate domain.

In order to predict the price of a given property, the buyer must have an idea about the final price of this property so that he would be able to negotiate the price and counter the strategy of the seller. Here, the use of the 3R model allows us to estimate the selling price of a property and this is in a context where information lacks. Tests are carried out on a base of 16 cases, and in order to check and test the model, we have taken another 4 cases to be estimated. We have come up with interesting results (Table 2) that establish the efficiency and the relevance of this approach. We notice that the graphs (Figure 12) representing the real and the predicted/estimated values develop the same way and that the difference between the two series is insignificant. Therefore, we can say that the 3R model adapts well to real estate negotiation.

5. Conclusion

To make problem resolution easier in a context where the case-base is incomplete, i.e. in a dynamic environment, and to make an incremental and progressive evolution of the base possible, we have proposed to use the CBR techniques to be able to adapt easily to the environment.

In this work of research, we have conceived a model to predict the selling prices in a negotiation. This model is a strategy that allows us to anticipate the seller's behaviour in the course of a negotiation and monitor his strategy. This model based on the CBR is a reduced form of the classical CBR given that it is set on three stages -hence the name the 3R model, namely: "retrieve", "reuse" and "retain":

Retrieve: this step enables us to carry out research to determine the optimal weights in order to use them to pick up the most similar case.

Reuse: this one allows us to make use of the solution of the similar case to calculate the solution of the target case in one go without adaptation and by applying a mathematical formula.

Retain: this step allows the storage of the new resolved cases and thus get the base to develop progressively.

This 3R model is our contribution in this work of research. Applying the model to the real estate domain entails interesting results.

A fundamental aspect of our contribution is concerned with the original combination of the CBR technique and the mathematical tools for the resolution of the problem and for decision making, and this during the « retrieve » and « reuse » stages.

At the level of the retrieve stage, our contribution consists in splitting this stage into two phases, namely: "retrieve 1" and "retrieve 2", and the automatic determination of the weights.

In the "reuse" stage, our contribution is the simple use of the solution of the similar case to reach the target solution. So there is no iterative cycle of modification, checking and testing to confirm the target solution. It is just the implementation of a mathematical formula that allows us to reach the target solution. This way, we discard human intervention to validate the final solution and the cyclical stage of adaptation and make the "reuse" stage automatic through the direct use of the solution of the similar case: a logical result and final solution that is valid.

An important dimension of our work is the move towards the use of intelligent agents in EC to improve online negotiation. We apply it to real situations by integrating it in negotiation scenarios with the actors being involved, virtual enterprises and auction sites. Ultimately, to ensure a wide recognition of our model of prediction, it is necessary to provide empirical evidence using bases containing a large number of cases and diverse domains.

References

[1] Aamodt, A. & Plaza, E. (1994). "Case-based reasoning: Foundational issues,

methodological Variations, and system approaches", Artificial Intelligence Communications, IOS Press, 7(1), 39-59.

[2] Aha, W. D. (1991). "Case-Based Learning Algorithms". *DARPA Case-Based Reasoning Workshop*, Morgan Kaufmann, LoaAtlos, CA.

[3] Althoff K. D. (2001). "Case-Based Reasoning". *In: S. K. Chang (Ed.), Handbook on Software Engineering and Knowledge Engineering. Vol. 1*, "Fundamentals", World Scientific, 549-588.

[4] Bergmann R., Breen S., Göker M., Manago M. & Wess S. (1999). "Developing industrial case-based reasoning applications". *LNAI*, 1612, Springer.

[5] Chavez, A. & Maes, P .(1996). "KASBAH: An Agent Marketplace for Buying and Selling Goods". *In 1st Intl. Conf. on Practical Applications of Intelligent Agents Technology*, London, UK, April.

[6] Darling, T. A. & Mumpower, J. L. (1990). Modeling Cognitive Influences on the Dynamics of Negotiations. *Proceedings of the Twenty-third Hawaii International Conference on System Sciences*, 22-32.

[7] Dworman, Garett, Steven Kimbrough, and James Laing. (1995). "On Automated Discovery of Models Using Genetic Programming in Game-Theoretic Contexts," http://opim.wharton.upenn.edu/ ~dworman/my-papers/HICSSGP6.ps, January 1995, *forthcoming in Journal of Management Information Systems, vol. 12(3)*, Winter 1996.

[8] Esyin, C. (2003). "Iterative Matching On Multidimensional Case-Based Reasoning System For Negotiation". *Master Thesis*, University of Malaya.

[9] Fisher, R. & Ury, W. (1981). Getting to Yes: Negotiating Agreement Without Giving. In, *Houghton Mifflin Co.* Boston, MA.

[10] Gulliver, P. H. (1979). Disputes and Negotiations: *A Cross Cultural Perspective*. New York: Academic Press.

[11] Hayes, C., Cunningham, P. & Doyle, M. (1998). "Distributed CBR using XML". In: *Workshop for Intelligent Systems and Electronic Commerce* (within German Conference on Artificial Intelligence (KI'98)) September 15-17.

[12] Holsapple, C. W., Lai, H. & Whinston, A. B. (1998). A Formal Basis for Negotiation Support System Research. *Group Decision and Negotiation*, *7(3)*, 192-202.

[13] Hu J., Reeves D. & Wong H. S. (1999). "Agent service for online auctions". *In Workshop on Artificial Intelligence in Electronic Commerce*, Menlo Park, CA.

[14] Kersten, G. E., Noronha S. J. & Teich J. (2000). "Are All E-Commerce Negotiations Auctions?", *Proceedings of the 4th International Conference on the Design of Cooperative Systems*, Sophia-Antipolis, France, 1-11, May.

[15] Kersten. G. E. (2004). E-negotiation Systems: Interaction of People and Technologies to Resolve Conflicts. INR08/04 UNESCAP *Third Annual Forum on Online Dispute Resolution Melbourn*, Australia.

[16] Kolodner J. L. (1993). *"Case-Based Reasoning"*. Morgan Kaufmann.

[17] Kraus, S., Sycara, K. & Evenchik, A. (1998). "Reaching Agreements through Argumentation: A Logical Model and Implementation", *Artificial Intelligence, 104(1-2)*, 1-69.

[18] Lai, H., Doong, H. S., Kao, C. C. & Kersten, G. E. (2006). "Understanding Behavior and Perception of Negotiators from Their Strategies". *Hawaii International Conference on System Science.*

[19] Lax, D. A. & Sebenius, J. K. (1986). "*The Manager as Negotiator: Bargaining for*

Cooperative and Competitive Gain". New York: The Free Press.

[20] Leake D. B. (1996). "CBR in context: The present and future; *in Leake, D. B. (editor) Case-Based Reasoning: Experiences, Lessons & Future Directions, American Association for Artificial Intelligence*, Menlo Park California, USA, 3-30.

[21] Maes, P., Guttman, R. H. & Moukas, A. G. (1999). "Agents that Buy and Sell: Transforming Commerce as we know it". *Communications of the ACM*, March.

[22] Oliver, Jim, R. (1996). "A Machine Learning Approach to Automated Negotiation and Prospects for Electronic Commerce," *http://opim.wharton.upenn.edu/~oliver2* 7/papers/jmis.ps, July 31.

[23] Pruitt, D. G. & Carnevale, P. J. (1993). *Negotiation in Social Conflict*. Buckingham: Open Univ. Press.

[24] Schaaf, M., Freßmann, A., Maximini, R., Bergmann, R., Tartakovski, A. & Radetzki, M. (2004). "Intelligent IP Retrieval Driven by Application Requirements", *Integration, the VLSI Journal, 37(4)*, 253-287.

[25] Schank, R. (1982). *"Dynamic Memory: A Theory of Reminding and Learning in Computer and People"*. Cambridge University Press.

[26] Su, S., Huang, C. & Hammer, J. (2000). *"A Replicable Web-Based Negotiation Server for E-Commerce"*, Proceedings of the Thirty -Third Hawaii International Conference on System Sciences (HICSS-33) - *Volume 8*, IEEE Computer Society, Hawaii, USA.

[27] Sun, Z., Finnie, G. & Weber, K. (2003). "A similarity-based theory of case-based reasoning-I", TR03-02. *School of Information Technology*, Bond University. 2003. Available at: http:// www. it. bond. edu. au/ publications/03TR/03-02.pdf.

[28] Tartakovski, A., Schaaf, M., Maximini, R. & Bergmann, R. (2004). "MINLP Based Retrieval of Generalized Cases", Proceedings of 7th European Conference, ECCBR 2004. In Peter Funk, and Pedro A. González Calero, editors, *Advances in Case-Based Reasoning*, LNAI 3155, pages 404-418, Madrid, Spain, August 2004, Springer Verlag, Berlin-Heidelberg.

[29] Thomas, K. W. (1976). Conflict and Conflict Management, In *Handbook of Industrial and Organizational Psychology*, ed. MD Dunnette, Chicago: Rand McNally, 889-935.

[30] Wettschereck, D. & Aha, D. W. (1995). "Weighting Features", *Proceedings of the First International Conference on Case-Based Reasoning*, Springer, New York.

[31] Wilke, W., Bergmann, R. & Wess, S. (1998). "Negotiation During Intelligent Sales Support with Case-Based Reasoning", Proceedings of the 6th German Workshop on Case-Based Reasoning (GWCBR'98), *http://wwwagr.informatik.uni-kl.de/index2.ht* ml?click=2.

[32] Wong, W. Y., Zhang, D. M. & Kara-Ali, M. (2000). "Negotiating With Experience", *Proceedings of the Knowledge-based Electronic Markets (KBEM'00)*, Austin TX, USA, 85.

[33] Wu, S., Kersten, G. & Benyoucef, M. (2006). "Web-based Negotiation Support Systems". *Proceedings of the Montreal Conference on e-Technologies*.

[34] Zhang, D. M. & Wong, W. Y. (2001). "A Web-based Negotiation Agent Using CBR". *West Yorkshire: Lecturer Notes in Artificial Intelligence*, *Vol. 2112*, 183-195.

[35] Zhang, D. M. & Wong, W. Y. (2000). "Using CBR for adaptive negotiation". In: Z., Shi, B., Faltings, & M. Musen, (eds): *Proc Conf on Intelligent Information Processing (IIP2000)*, Aug. 21-25, 2000 Beijing. 428-37.

In: Case-Based Reasoning: Processes Suitability...
Editors: Antonia M. Leeland
ISBN: 978-1-61728-352-9

Commentary

FUZZY SETS IN CASE-BASED REASONING

Michael Gr. Voskoglou*
Technological Educational Institute (T. E. I.), Patras, Greece

Case-Based Reasoning (CBR), as a general problem-solving methodology intended to cover a wide range of real-world applications, must face the challenge of dealing with uncertain, incomplete and vague information. Successfully deployed CBR systems are commonly integrated with some method for treating uncertainty, which is already inherent in the basic CBR hypothesis, demanding that similar problems have similar solutions. Correspondingly, recent years have witnessed an increased interest in formalizing parts of the CBR methodology within different frameworks of reasoning under uncertainty, and in building hybrid approaches by combining CBR with methods of uncertain and approximate reasoning. Fuzzy sets theory can be mentioned as a particularly interesting example. In fact, even though both CBR and fuzzy systems are intended to be cognitively more plausible approaches to reasoning and problem-solving, the two corresponding fields have emphasized different aspects that complement each other in a reasonable way. Thus, fuzzy set-based concepts and methods can support the various aspects of CBR, while on the other hand, ideas and techniques for CBR can contribute to fuzzy set-based approximate reasoning.

Recently we have constructed a fuzzy model to describe a CBR system by representing the main steps of the CBR process (retrieval, reuse, revision, retaining) as fuzzy subsets of the set, called U, of the linguistic labels of negligible, low, intermediate, high and complete success in each of the above steps. The total possibilistic uncertainty is used in our model as a measurement of the effectiveness of the CBR system under study in solving new related problems. Examples have also been produced to illustrate our results, but further study and research are needed for testing the usefulness of our model in practice.

* Corresponding author: e-mail: voskoglou@teipat.gr ; mvosk@tellas.gr.

INDEX

A

B

C

D

E

F

G

H

I

S

T

U

V

W

X

Y